REALISM IN MATHEMATICS

REALISM IN MATHEMATICS

PENELOPE MADDY

CLARENDON PRESS · OXFORD

Oxford University Press, Walton Street, Oxford OX2 6DP
Oxford New York Toronto
Delhi Bombay Calcutta Madras Karachi
Petaling Jaya Singapore Hong Kong Tokyo
Nairobi Dar es Salaam Cape Town
Melbourne Auckland
and associated companies in
Berlin Ibadan

Oxford is a trade mark of Oxford University Press

Published in the United States
by Oxford University Press, New York

First published 1990
Hardback reprinted 1992

First issued in paperback with corrections 1992

British Library Cataloguing in Publication Data
Maddy, Penelope
Realism in mathematics.
1. Mathematics. Philosophical perspectives, history
I. Title
510.1
ISBN 0–19–824035–X

Library of Congress Cataloging in Publication Data
Maddy, Penelope.
Realism in mathematics / Penelope Maddy.
p. cm.
Includes bibliographical references.
1. Mathematics—Philosophy. I. Title.
QA8.4.M33 1990 510′.1—dc20 89–49346
ISBN 0–19–824035–X

Set by Pentacor PLC, *High Wycombe, Bucks*
Printed in Great Britain by
Biddles Ltd, Guildford & King's Lynn

For
Dick and Steve

PREFACE

THE philosophy of mathematics is a borderline discipline, of fundamental importance to both mathematics and philosophy. Despite this, one finds surprisingly little co-operation between philosophers and mathematicians engaged in its pursuit; more often, widespread disregard and misunderstanding are broken only by alarming pockets of outright antagonism. (The glib and dismissive formalism of many mathematicians is offset by the arrogance of those philosophers who suppose they can know what mathematical objects are without knowing what mathematics says they are.) This might not matter much in another age, but it does today, when the most pressing foundational problems are unlikely to be answered without a concerted co-operative effort. I have tried in this book to do justice to the concerns of both parties, to present the background, the issues, the proposed solutions on a neutral ground where the two sides can meet for productive debate.

For this reason, I've aimed for a presentation accessible to both non-philosophical mathematicians and non-mathematical philosophers and, if I've succeeded, students and interested amateurs should also be served. As far as I can judge, very little philosophical training or background is presupposed here. Mathematical prerequisites are more difficult to avoid, owing to the relentlessly cumulative nature of the discipline, but I've tried to keep them to a minimum. Some familiarity with the calculus and its foundations would be helpful, though surely not necessary. And the relevant set theoretic concepts are referenced to Enderton's excellent introductory textbook (see his (1977)), for the benefit of those innocent of that subject.

The central theme of the book is the delineation and defence of a version of realism in mathematics called 'set theoretic realism'. In this, my deep and obvious debt is to the writings of the great mathematical realists of our day: Kurt Gödel, W. V. O. Quine, and

Hilary Putnam (in the early 1970s). More personally, I have learned most from John Burgess, Paul Benacerraf, Hartry Field, and Tony Martin. After these, it would be impossible to mention everyone, but I can't overlook the forceful criticisms of Charles Chihara, the insightful comments of Anil Gupta, and the generous correspondence, assistance, and advice of Philip Kitcher and Michael Resnik. Most recently, Burgess, Field, Lila Luce, Colin McLarty, Martin, Alan Nelson, Resnik, Stewart Shapiro, Mark Wilson, and Peter Woodruff have all done me the service of reading and reacting to drafts of various parts of the manuscript. (Naturally, the remaining errors and oversights should be charged to my shortcomings rather than to their negligence.) And finally, what I owe to my long-time companion Steve Maddy is too complex and varied to be summarized here. I am grateful to all these people and offer my heartfelt thanks. Also to Angela Blackburn and Frances Morphy of Oxford University Press.

Much of this book is based on a series of articles (Maddy 1980, 1981, 1984*a*, 1988*a*,*b*, forthcoming *a*, *b*) the preparation of which was supported, at various times, by the American Association of University Women, the University of Notre Dame, the National Endowment for the Humanities, the National Science Foundation, and the University of Illinois at Chicago. The original publishers kindly granted advance permission to reproduce material from these pieces; in the end, only parts of (forthcoming *a*) (in chapter 5, sections 1 and 2) and (forthcoming *b*) (in chapter 1, section 4) actually survived, so I am particularly obliged to Kluwer Academic Publishers and the Association for Symbolic Logic. Preparation of the final draft was supported by National Science Foundation Grant DIR-8807103, a University of California President's Research Fellowship in the Humanities, and the University of California at Irvine. The help of all these institutions is hereby gratefully acknowledged.

Finally, I feel compelled to add a personal note on sexist language. Some years ago, when I first introduced the ideas behind set theory realism, constructions like 'the set theoretic realist thinks *his* entities . . .' struck me as amusing, but since then I've discovered that some readers and editors are legitimately disapproving of this usage. Of the many alternatives available, I've chosen one that does the least violence to the standard rhythm, that is, the use of 'she' and 'her' in place of 'he' and 'his' in neutral contexts. Some might find this just as politically incorrect as the automatic use of the masculine, but I sincerely doubt that phrasing like 'when the mathematician proves a theorem, *she* . . .' makes anyone tend to forget that some mathematicians are men. So I'll stick with this policy. To those who find it distracting, I apologize; this is not, after all, a political treatise. At least you have my reasons.

P.M.

Irvine, California
June 1989

CONTENTS

1
REALISM

1. Pre-theoretic realism

Of the many odd and various things we believe, few are believed more confidently than the truths of simple mathematics. When asked for an example of a thoroughly dependable fact, many will turn from common sense—'after all, they used to think humans couldn't fly'—from science—'the sun has risen every day so far, but it might fail us tomorrow'—to the security of arithmetic—'but 2 plus 2 is surely 4'.

Yet if mathematical facts are facts, they must be facts about something; if mathematical truths are true, something must make them true. Thus arises the first important question: what is mathematics about? If 2 plus 2 is so definitely 4, what is it that makes this so?

The guileless answer is that $2 + 2 = 4$ is a fact about numbers, that there are things called '2' and '4', and an operation called 'plus', and that the result of applying that operation to 2 and itself is 4. '$2 + 2 = 4$' is true because the things it's about stand in the relation it claims they do. This sort of thinking extends easily to other parts of mathematics: geometry is the study of triangles and spheres; it is the properties of these things that make the statements of geometry true or false; and so on. A view of this sort is often called 'realism'.

Mathematicians, though privy to a wider range of mathematical truths than most of us, often incline to agree with unsullied common sense on the nature of those truths. They see themselves and their colleagues as investigators uncovering the properties of various fascinating districts of mathematical reality: number theorists study the integers, geometers study certain well-behaved spaces, group theorists study groups, set theorists sets, and so on. The very experience of doing mathematics is felt by many to support this position:

The main point in favor of the realistic approach to mathematics is the instinctive certainty of most everybody who has ever tried to solve a problem that he is thinking about 'real objects', whether they are sets, numbers, or whatever . . . (Moschovakis (1980), 605)

Realism, then (at first approximation), is the view that mathematics is the science of numbers, sets, functions, etc., just as physical science is the study of ordinary physical objects, astronomical bodies, subatomic particles, and so on. That is, mathematics is about these things, and the way these things are is what makes mathematical statements true or false. This seems a simple and straightforward view. Why should anyone think otherwise?

Alas, when further questions are posed, as they must be, embarrassments arise. What sort of things are numbers, sets, functions, triangles, groups, spaces? Where are they? The standard answer is that they are abstract objects, as opposed to the concrete objects of physical science, and as such, that they are without location in space and time. But this standard answer provokes further, more troubling questions. Our current psychological theory gives the beginnings of a convincing portrait of ourselves as knowers, but it contains no chapter on how we might come to know about things so irrevocably remote from our cognitive machinery. Our knowledge of the physical world, enshrined in the sciences to which realism compares mathematics, begins in simple sense perception. But mathematicians don't, indeed can't, observe their abstract objects in this sense. How, then, can we know any mathematics; how can we even succeed in discussing this remote mathematical realm?

Many mathematicians, faced with these awkward questions about what mathematical things are and how we can know about them, react by retreating from realism, denying that mathematical statements are about anything, even denying that they are true: 'we believe in the reality of mathematics, but of course when philosophers attack us with their paradoxes we rush to hide behind formalism and say "Mathematics is just a combination of meaningless symbols" . . .'.[1] This formalist position—that mathematics is just a game with symbols—faces formidable obstacles of its own, which I'll touch on below, but even without these, many mathematicians find it involving them in an uncomfortable form of

[1] Dieudonne, as quoted in Davis and Hersh (1981), 321.

double-think. The same writer continues: 'Finally we are left in peace to go back to our mathematics and do it as we have always done, with the feeling each mathematician has that he is working on something real' (Davis and Hersh (1981), 321). Two more mathematicians summarize:

> the typical working mathematician is a [realist] on weekdays and a formalist on Sundays. That is, when he is doing mathematics he is convinced that he is dealing with an objective reality whose properties he is attempting to determine. But then, when challenged to give a philosophical account of this reality, he finds it easiest to pretend that he does not believe in it after all. (Davis and Hersh (1981), 321)

Yet this occasional inauthenticity is perhaps less troubling to the practising mathematician than the daunting requirements of a legitimate realist philosophy:

> Nevertheless, most attempts to turn these strong [realist] feelings into a coherent foundation for mathematics invariably lead to vague discussions of 'existence of abstract notions' which are quite repugnant to a mathematician . . . Contrast with this the relative ease with which formalism can be explained in a precise, elegant and self-consistent manner and you will have the main reason why most mathematicians claim to be formalists (when pressed) while they spend their working hours behaving as if they were completely unabashed realists. (Moschovakis (1980), 605–6)

Mathematicians, after all, have their mathematics to do, and they do it splendidly. Dispositionally suited to a subject in which precisely stated theorems are conclusively proved, they might be expected to prefer a simple and elegant, if ultimately unsatisfying, philosophical position to one that demands the sort of metaphysical and epistemological rough-and-tumble a full-blown realism would require. And it makes no difference to their practice, as long as double-think is acceptable.

But to the philosopher, double-think is not acceptable. If the very experience of doing mathematics, and other factors, soon to be discussed, favour realism, the philosopher of mathematics must either produce a suitable philosophical version of that position, or explain away, convincingly, its attractions. My goal here will be to do the first, to develop and defend a version of the mathematician's pre-philosophical attitude.

Rather than attempt to treat all of mathematics, to bring the project down to more manageable size, I'll concentrate here on the mathematical theory of sets.[2] I've made this choice for several reasons, among them the fact that, in some sense, set theory forms a foundation for the rest of mathematics. Technically, this means that any object of mathematical study can be taken to be a set, and that the standard, classical theorems about it can then be proved from the axioms of set theory.[3]

Striking as this technical fact may be, the average algebraist or geometer loses little time over set theory. But this doesn't mean that set theory has no practical relevance to these subjects. When mathematicians from a field outside set theory are unusually frustrated by some recalcitrant open problem, the question arises whether its solution might require some strong assumption heretofore unfamiliar within that field. At this point, practitioners fall back on the idea that the objects of their study are ultimately sets and ask, within set theory, whether more esoteric axioms or principles might be relevant. Given that the customary axioms of set theory don't even settle all questions about sets,[4] it might even turn out that this particular open problem is unsolvable on the basis of these most basic mathematical assumptions, that entirely new set theoretic assumptions must be invoked.[5] In this sense, then, set theory is the ultimate court of appeal on questions of what mathematical things there are, that is to say, on what philosophers call the 'ontology' of mathematics.[6]

Philosophically, however, this ontological reduction of mathematics to set theory has sometimes been taken to have more dramatic consequences, for example that the entire philosophical foundation of any branch of mathematics is reducible to that of set theory. In this sense, comparable to implausibly strong versions of

[2] A set is a collection of objects. Among the many good introductions to the mathematical theory of these simple entities, I recommend Enderton (1977).

[3] See e.g. the reduction of arithmetic and real number theory to set theory in Enderton (1977), chs. 4 and 5. There are some exceptions to the rule that all mathematical objects can be thought of as sets—e.g. proper classes and large categories—but I will ignore these cases for the time being.

[4] Some details and philosophical consequences of this situation are the subject of ch. 4.

[5] Eklof and Mekler (forthcoming) give a survey of algebraic examples, and Moschovakis (1980) does the same for parts of analysis.

[6] In philosophical parlance, 'ontology', the study of what there is, is opposed to 'epistemology', the study of how we come to know what we do about the world. I will use the word 'metaphysics' more or less as a synonym for 'ontology'.

the thesis that physics is basic to the natural sciences,[7] I think the claim that set theory is foundational cannot be correct. Even if the objects of, say, algebra are ultimately sets, set theory itself does not call attention to their algebraic properties, nor are its methods suitable for approaching algebraic concerns. We shouldn't expect the methodology or epistemology of algebra to be identical to that of set theory any more than we expect the biologist's or the botanist's basic notions and techniques to be identical to those of the physicist. But again, this methodological independence of the branches of mathematics from set theory does not mean there must be mathematical entities other than sets any more than the methodological independence of psychology or chemistry from physics means there must be non-physical minds or chemistons.[8]

But little hangs on this assessment of the nature of set theory's foundational role. Even if set theory is no more than one among many branches of mathematics, it is deserving of philosophical scrutiny. Indeed, even as one branch among many, contemporary set theory is of special philosophical interest because it throws into clear relief a difficult and important philosophical problem that challenges many traditional attitudes toward mathematics in general. I will raise this problem in Chapter 4.

Finally, it is impossible to divorce set theory from its attendant disciplines of number theory and analysis. These two fields and their relationship to the theory of sets will form a recurring theme in what follows, especially in Chapters 3 and 4.

2. Realism in philosophy

So far, I've been using the key term 'realism' loosely, without clear definition. This may do in pre-philosophical discussion, but from

[7] This view is called 'physicalism'. I'll come back to it in ch. 5, sect. 1, below.

[8] There was a time when the peculiarities of biological science led practitioners to vitalism, the assumption that a living organism contains a non-physical component or aspect for whose behaviour no physical account can be given. Nowadays, this idea is discredited—simply because it proved scientifically sterile—and, as far as I know, no one ever urged the acceptance of 'chemistons'. Today, psychology is the special science that most often lays claim to a non-physical subject matter, but as suggested in the text, it seems to me that a purely physical ontology is compatible with the most extreme methodological independence. For discussion, see Fodor (1975), 9–26.

now on I will try to be more precise. This doesn't mean I'll succeed in defining the term exactly, but at least I'll narrow the field somewhat, I hope helpfully. Let me begin by reviewing some traditional uses of the term in philosophy.

One of the most basic ontological debates in philosophy concerns the existence of what common sense takes to be the fundamental furniture of the world: stones and trees, tables and chairs, medium-sized physical objects. Realism in this context, often called 'common-sense realism', affirms that these familiar macroscopic things do in fact exist. But it is not enough for the realist to insist that there are stones and trees and such like, for in this much the idealist could agree, all the while assuming that a stone is a mental construct of some sort, say a bundle of experiences. However, such an idealist, like the Bishop Berkeley, will have serious trouble agreeing with the realist that stones can exist without being perceived.[9] Thus the common-sense realist can state her position in a way that rules out idealism by claiming that stones etc. exist, and that their existence is non-mental, that they are as they are independently of our ability to know about them, that their existence is, in a word, 'objective'.

A more recent opponent of the common-sense realist uses a more devious technique.[10] The phenomenalist hopes to say exactly what the realist says while systematically reinterpreting each and every physical object claim into a statement about what she calls 'sense data', or really, into statements about possible sense data. For example, my overcoat exists in the closet though unperceived because part of the translation of 'the overcoat is in the closet' is something like 'if I were in the closet and the light were on, then I'd have an overcoat-like experience', which is, presumably, true. Physical objects are not taken to consist of ideas, as with Berkeley, but physical object statements are taken to mean something other than what we ordinarily take them to mean.

[9] Berkeley's notorious solution was to suppose that God is perceiving the object even when we aren't; indeed he uses this as a novel argument for the existence of God. See, e.g. Berkeley (1713), 211–13, 230–1. It's worth noting, however, that in earlier work, Berkeley (1710 §§ 3, 58–9) includes a 'counterfactual' analysis that prefigures the Millian phenomenalism described in the next paragraph.

[10] This idea took shape in Mill (1865), ch. 11 and its appendix, and was developed in the form described here by the logical positivists. See Ayer (1946), 63–8.

This ambitious programme was a complete failure for a number of nagging reasons, only the first of which is our seeming inability to specify the required sense datum—the overcoat-like experience —without reference to the overcoat itself.[11] But whatever the failings of phenomenalism, the attempt itself shows that the realist, to state her position completely, must also rule out such unintended misinterpretations of common-sense statements; she must insist that these statements be taken 'at face value'. Because it is hard to say exactly what this comes to, apart from repeating that it rules out phenomenalism, realism is in some ways more difficult to state than its particular rivals. In any case, we can be sure that common-sense realism is opposed to both idealism and phenomenalism.

Our discussion so far has centred on the problem of stating common-sense realism; we must now ask why we should believe it. The failure of heroic philosophical alternatives like idealism and phenomenalism is some reassurance, but we would like a positive argument. Admittedly, we find it difficult *not* to believe in ordinary physical objects; of his own philosophical scepticism, the great David Hume writes:

> since reason is incapable of dispelling these clouds, nature herself suffices to that purpose . . . I dine, I play a game of backgammon, I converse, and am merry with my friends; and when after three or four hours' amusement, I would return to these speculations, they appear so cold, and strained, and ridiculous, that I cannot find in my heart to enter into them any farther. (Hume (1739), 548–9)

But even if common-sense realism is psychologically inevitable, we should still ask after its justification.

The reply given by many contemporary philosophers is simply that the existence of ordinary things provides the best account of our experience of the world. In his landmark essay on ontology, 'On what there is', W. V. O. Quine puts the point this way:

> By bringing together scattered sense events and treating them as perceptions of one object, we reduce the complexity of our stream of experience to a manageable conceptual simplicity. . . . we associate an earlier and a later round sensum with the same so-called penny, or with two different so-called pennies, in obedience to the demands of maximum simplicity in our total world-picture. (Quine (1948), 17)

[11] See Urmson (1956), ch. 10, for a survey of this and other difficulties.

Now we can hardly be said to make an explicit inference from purely experiential statements to physical object statements that account for them, because (as noted in connection with phenomenalism) we have no independent language of experience. What actually happens is a developing neurological mediation between purely sensory inputs and our primitive beliefs about physical objects.[12] The justificatory inference comes later, when we argue that the best explanation of our stubborn belief in physical objects is that they do exist and that our beliefs about them are brought about in various dependable ways, for example by light bouncing off their surfaces on to our retinas etc. Thus the assumption of objectively existing, medium-sized physical objects plays an indispensable role in our best account of experience.

But, one might object, didn't the gods of Homer provide the Greeks with an explanation of their experience? Here Quine points to an important difference:

> For my part I do, qua lay physicist, believe in physical objects and not in Homer's gods; and I consider it a scientific error to believe otherwise. . . . The myth of physical objects is epistemologically superior to most in that it has proved more efficacious than other myths as a device for working a manageable structure into the flux of experience. (Quine (1951), 44)

Physical objects, not Homer's gods, form part of our best scientific theory of the world, and for that reason, our belief in the former, but not the latter, is justified.

Notice, however, that this sort of answer will not satisfy the philosophical sceptic who calls all our belief-forming techniques, including those of natural science, into question. René Descartes, for example, was well aware that science presupposed an objective external world, but he wanted a justification for science itself. How can we know, Descartes asked, that the scientific world-view is correct? How do we know our senses aren't deceiving us? How do

[12] For more on this, see ch. 2, sect. 2, below. Students of Quine may detect a tension between my position in this paragraph and such Quinean remarks as 'From among the various conceptual schemes best suited to these various pursuits, one—the phenomenalistic—claims epistemological priority' ((1948), p. 19). Here and in what follows, I will ignore this lingering trace of positivism in the master. In fact, there is no phenomenalistic language or theory, and a good scientific explanation must do more than accurately predict sense experiences. (Cf. Putnam (1971), 355–6.)

we know we aren't dreaming? How do we know there is no Evil Demon systematically making it appear to us as if the world is as we think it is?[13]

These Cartesian challenges depend on a conception of epistemology as an a priori[14] study of knowledge and justification, a study above, beyond, outside, indeed prior to, natural science, a study whose aim is to establish that science on a firm footing. One might think that the justificatory practices of science itself are the best we have, but classical epistemology appeals to higher canons of pure reason. Unfortunately, its attempts to reconstruct natural science on an a priori, philosophically justified foundation have all failed, beginning with Descartes's own effort.[15]

In light of this history, Quine suggests a radically different approach to epistemology. Our best understanding of the world, after all, is our current scientific theory, so by what better canons can we hope to judge our epistemological claims than by scientific ones? The study of knowledge, then, becomes part of our scientific study of the world, rather than an ill-defined, pre-scientific enterprise: 'Epistemology ... simply falls into place as a chapter of psychology and hence of natural science. It studies a natural phenomenon, viz., a physical human subject' (Quine (1969*b*), 82). Standing within our own best theory of the world—what better perspective could we have?—we ask how human subjects like ourselves are able to form reliable beliefs about the world as our theory tells us it is. This descriptive and explanatory project is called 'epistemology naturalized'.

Thus science is used to justify science, but this circle is not vicious once we give up the classical project of founding science on something more dependable than itself. From naturalized perspective, there is no point of view prior or superior to that of natural science, and Quine's argument for common-sense realism becomes perfectly reasonable: the assumption of physical objects is part of our best theory, and being part of our best theory is the best justification a belief can have.

[13] Descartes (1641), esp. Meditation One.

[14] 'A priori' means prior to experience, as opposed to 'a posteriori'.

[15] Descartes argued that our perceptions are reliable because God is no deceiver. Serious objections to this approach arose immediately; see his 'Objections and replies', published as an appendix to his *Meditations*. A more recent failed effort to found science is that of the positivists. See Quine's discussion (1969*b*).

A second form of philosophical realism concerns itself with the more esoteric objects of science, with unobservable theoretical entities like electrons, genes, and quarks. Here the scientific realist asserts that our belief in such things is justified, at least to the extent that theories involving them provide us with the best explanation we have for the behaviour of observable objects. Once again, however, it is not enough for the realist to say just this. While there can be no idealist here, analogous to the Berkeleian, insisting that electrons are just bundles of sensory experiences, there is a position analogous to phenomenalism: instead of translating talk about medium-sized physical objects into talk about what sensory experiences would occur under what circumstances, the operationalist would have us translate talk of unobservable theoretical entities into talk about how observables would behave under which circumstances.[16] Thus, for example, part of the translation of 'there's a quark here' might be 'if we set up a cloud chamber, we'd get this kind of track' and 'if we prepared a photographic emulsion, we'd see this kind of trace' and 'if we had a scintillation counter, we'd get this type of signal'. This project failed as resoundingly as phenomenalism,[17] and for some of the same reasons, but again, its very existence shows that along with asserting the existence of those unobservables presupposed by our best theory, the scientific realist must also insist that this assertion be taken 'at face value'.

The scientific realist's most conspicuous opponent is the instrumentalist, who holds that unobservables are a mere 'useful fiction' that helps us predict the behaviour of the observable. Thus the instrumentalist denies just what the scientific realist asserts—that there are electrons etc.—but continues to use the same theories the realist does to predict the behaviour of observables. For a practising scientist, instrumentalism would seem as dramatic a form of double-think as the duplicitous mathematical formalism described earlier. But there are worse problems than inauthenticity. For example, the distinction between theoretical and observational turns out to be devilishly hard to draw.[18] And even if it could be

[16] The classic statement of operationalism is Bridgman (1927). It is criticized by Hempel (1954), but the logical positivists found more subtle forms (e.g. Carnap (1936/7)).

[17] One of its many critics is Putnam (1962). Others are Maxwell (1962) and Achinstein (1965).

[18] See the papers cited in the previous footnote.

drawn, it is unclear why the difference between being humanly observable and not should have such profound metaphysical consequences.[19] Finally, the instrumentalistic scientist, happily using the false premisses of theoretical science to derive purportedly true conclusions about observables, provides us no explanation of why a batch of false claims should be so dependable.

So, once again, as in the case of common-sense realism, the failure of the opposition provides some negative support for scientific realism. The positive argument is also analogous.[20] After the above-quoted defence of common sense realism, Quine continues:

> Positing does not stop with macroscopic physical objects. Objects at the atomic level are posited to make the laws of macroscopic objects, and ultimately the laws of experience, simpler and more manageable . . . Science is a continuation of common sense, and it continues the common-sense expedient of swelling ontology to simplify theory. . . . Epistemologically these are myths on the same footing with physical objects and gods, neither better nor worse except for the differences in the degree to which they expedite our dealings with sense experiences. (Quine (1951), 44–5)

From the point of view of epistemology naturalized, what better justification could we have to believe in the most well-confirmed posits of our best scientific theory than the fact that they *are* the most well-confirmed posits of our best scientific theory?

The final realist/anti-realist controversy I want to consider is in fact the oldest debate in which the term 'realism' arises. We common-sense realists all agree that there are many red things—red roses, red houses, red sunsets—but the ancient question is whether or not there is also, over and above this lot of particular red things, a further thing they share, namely, redness. Such an additional thing—redness—would be a universal. The most basic difference between particulars and universals is that a universal can be present ('realized', 'instantiated', 'exemplified') in more than one place at a

[19] See Devitt (1984), §§ 7.1, 7.6, 8.5.

[20] The analogy is not perfect because with physical objects, unlike theoretical entities, we believe in them from the start; we are never in the position of deciding whether or not to begin believing in them. My point is that the form of the justification for the belief, however it is arrived at, is the same in both cases.

time, while a particular cannot.[21] Plato originated the most dramatic version of realism about universals in his spectacular theory of Forms: Redness, Equality, Beauty, and so on, are perfect, eternal, unchanging Forms; they exist outside of time and space; we know them by means of the non-sensory intellect; ordinary physical properties, perceived by the usual senses, are but pale and imperfect copies.[22] Aristotle, Plato's student, took direct aim at the more bizarre elements of this view, and defended a more modest form of realism, according to which universals exist only in those things that exemplify them.[23] Their opponent is the nominalist, like John Locke, who holds that there is nothing over and above particulars.[24]

One classical argument for realism about universals, the One over Many, survives in the modern debate. David Armstrong, its most vocal contemporary advocate, puts it like this: 'Its premise is that many different particulars can all have what appears to be the same nature' (Armstrong (1978), p. xiii); 'I would . . . draw the conclusion that, as a result, there is a *prima facie* case for postulating universals' (Armstrong (1980), 440–1). A similar argument can be given in linguistic form, arguing for example that a universal redness must exist since the predicate 'redness' is meaningful. Most contemporary thinkers, including Armstrong, reject this second form of the argument.[25] To see why, consider a scientific universal like 'being at a temperature of 32 degrees Fahrenheit'. Scientists tell us that this is the same universal, or property, to use a more natural-sounding word, as 'having such-and-such mean kinetic energy'.[26] But these two predicates have very different meanings. So properties are not meanings. They are

[21] A particular need not be spatially or temporally continuous—my copy of *Principia Mathematica* is in three volumes, on two different shelves—but even then, it is only part of the particular that is present in each of the disparate locations. A universal is understood to be fully present in each of its instances.

[22] The standard reference for Plato's theory is his *Republic*, chs. 5–7. Wedberg (1955), ch. 3, gives a useful summary. He cites the later *Timaeus* (37 D–38 A) as defining the sense in which Forms are outside time. A metaphorical passage in *Phaedrus* (247) declares their location as 'the heaven which is above the heavens', indicating they are not spatial, and Aristotle's commentary confirms this: 'the Forms are not outside, because they are nowhere' (*Physics*, 203^{a}8).

[23] See his *Metaphysics*, bk. 1, § 9 for criticisms of Plato, and Categories 2, for his own view.

[24] See Locke (1690), bk. 3, ch. 3, § 1.

[25] See Armstrong (1978), pt. 4. See also Putnam (1970), § 1.

[26] Wilson (1985) argues that this frequently cited example is more complex than philosophers ordinarily appreciate, but I don't think this important observation affects the point at issue here.

individuated by scientific tests, such as playing the same causal role, rather than by synonymy of predicates.

But even Armstrong's preferred form of the One over Many argument has been severely criticized from a number of different perspectives.[27] One stunningly simple counter-argument, Quine's, goes like this. We want to say that Ted and Ed are white dogs. This is supposed to commit us to the universal 'whiteness'. But for 'Ted and Ed are white dogs' to be true, all that is required is that there be a white dog named 'Ted' and a white dog named 'Ed'; no 'whiteness' or even 'dogness' is necessary. If there is more than this to the One over Many argument, the realist owes an account of what that is. If not, the realist needs some other support for the existence of these universals.

Contemporary thinkers have proposed a more modern argument, modern in the sense that it partakes of the 'naturalizing' tendency identified above in recent epistemology. Notice first that the One over Many is presented as an a priori philosophical argument for an ontological conclusion: there are universals. Now epistemology naturalized, as described above, has renounced the classical claim to a philosophical perspective superior to that of natural science. As a result, the Cartesian demand for certainty beyond the scientific was also rejected. From this new naturalized perspective, the considered judgement of science is the best justification we can have. Finally, we've seen that this shift in epistemological thinking produces a corresponding shift in ontological thinking; for example, despite philosophical qualms about unobservable entities, we should admit they exist if our best science tells us they do. The moral for the defender of universals is clear: to show that there are universals, don't try to give a pre-scientific philosophical argument; just show that our best scientific theory cannot do without them.

Much of the current debate takes this form.[28] The question at issue is whether intensional entities like universals are needed, or whether science can get by on extensional entities like sets.[29] Most

[27] See Quine (1948), Devitt (1980), Lewis (1983).

[28] With Putnam (1970) and Wilson (1985; and forthcoming) on the positive, Quine (1948; 1980*b*) on the negative.

[29] Universals are intensional because two of them can apply to the same particulars without being identical, for example, 'human being' and 'featherless biped'. Sets, by contrast, are extensional; two sets with the same members are identical.

will grant that sets are more promiscuous than universals; random elements can be gathered into a set even if they have no property in common. Furthermore, science seems to need a distinction between random collections and 'natural' ones; when we notice that all the ravens we've examined are black, we conclude that all ravens belong to the set of black things, not that all ravens belong to the set of things that are either black or not examined by us. The open question is whether the nominalist can deal with this distinction between natural and unnatural collections without appealing to universals.[30]

To summarize, then, to be a realist about medium-sized physical objects, the theoretical posits of science, or universals, is to hold that these entities exist, that they do so objectively—they are not mental entities, and they have the properties they do independently of our language, concepts, theories, and of our cognitive apparatus in general—and to resist various efforts—phenomenalism, operationalism—to reinterpret these claims. And, in the naturalized spirit, the realist assumes that the most strongly held of our current theoretical beliefs are probably at least approximately correct accounts of what these things are like. Beyond what's sketched above, I will pause no further over arguments for or against these three forms of philosophical realism,[31] but I will take something of a stand.

Most of what follows will presuppose both common-sense and scientific realism.[32] Indeed, as will come out below, the debate about the existence and nature of mathematical entities is almost always posed by comparing them with medium-sized physical objects and/or theoretical entities; the philosopher's temptation is to embrace common-sense and scientific realism while rejecting mathematical realism.[33] I want to remark, however, that I don't

[30] Lewis (1983) provides a useful survey of the debate.

[31] Devitt (1984) provides a useful compendium of arguments for common-sense and scientific realism.

[32] Some argue that common-sense and scientific realism are incompatible, because physics reveals that medium-sized physical objects are quite different from our common-sense conception. But showing we are often wrong about stones and tables is not the same as showing that these things don't exist. I see no problem in allowing that science can correct common sense. Devitt (1984, § 5.10) sketches a position of this sort.

[33] See e.g. Putnam (1975*b*), 74.

think the rough-and-ready mathematical realism introduced in the previous section stands or falls with these other realisms. If its central tenet is that mathematics is as objective a science as astronomy, physics, biology, etc., then this might remain true even if those natural sciences turn out not to be as objective as the realist thinks they are. In the long run, I'm much more interested in blocking a serious metaphysical or epistemological disanalogy between mathematics and natural science than I am in maintaining a strict realism about either.[34]

Finally, because I think the issues involved are not nearly as clear, I will remain officially neutral on universals. The question will arise (in Chapter 2, section 4; Chapter 3, section 2; and Chapter 5, sections 2 and 3) but I think nothing I say will hang on the ultimate resolution of the metaphysical debate outlined above. The problem in question is quite general; it is no more a problem for mathematics than it is for the rest of science.[35]

3. Realism and truth

In recent years, many philosophers have come to think that realism should be understood, not as a claim about what there is, but as a claim about semantics.[36] Whether one is a realist or not—about the objects of common sense, theoretical entities, universals, or mathematical objects—is said to depend on what one takes to be the conditions for the truth or falsity of the corresponding statements.

Now my own pre-philosophical statements about mathematical realism do involve what sounds like a semantic element: I claimed that mathematics is about numbers, sets, functions, etc., and that the way these things are is what makes mathematical statements true or false. This sort of talk can be read as espousing a

[34] For example, my argument against Wittgenstein (Maddy (1986)) takes the form: even if his general anti-realism is correct, still his strong maths/science disanalogy need not be accepted.

[35] Unless, of course, all mathematical entities *are* universals. A version of this view is considered in ch. 5, sect. 3, below.

[36] The influence here is Dummett's; see Dummett (1978), introd. and chs. 1, 10, and 14, and (1977), ch. 7. Devitt (1984) gives a more complete discussion of the relationships between realism and semantics, and his ch. 12 takes up Dummett's position in particular.

correspondence theory of truth, according to which the truth of a sentence depends partly on the structure of the sentence, partly on the relations between the parts of the sentence and extra-linguistic reality,[37] and partly on the nature of that extra-linguistic reality. One definitive aspect of correspondence theories is that what it takes for a sentence to be true might well transcend what we are able to know.

Semantic anti-realists, by contrast, want to identify the truth conditions of a sentence with something closer to our abilities to know, with that which justifies the assertion of the sentence, with some version of its 'verification conditions'. Notice that phenomenalism could be reinterpreted this way, as the claim that the truth of 'my overcoat is in the closet' reduces to the truth of various counterfactual conditionals like the one about what experiences I'd have if I were in the closet with the light on. Moves to verificationism are variously motivated—by the hope of avoiding scepticism, by the desire to eliminate metaphysics, by a disbelief in the objective reality of the entities in question, by attention to purported facts of language learning, by scepticism about the notion of correspondence truth itself.[38]

Thus these semantic thinkers identify realism about a certain range of entities with a correspondence theory of truth for sentences concerning those entities, and likewise anti-realism with verificationism. Given our previous characterization of realism as a position on what there is, such an identification seems wrong-headed. For example, an idealist like Berkeley could embrace a correspondence theory; for him, the extra-linguistic reality that makes ordinary physical object statements either true or false consists of bundles of experiences. Being a correspondence theorist doesn't make him a realist in our sense because his objects aren't objective.

On the other hand, our realist thinks her entities do exist objectively, which includes the belief that they exist and are as they are independently of our abilities to know about them. She holds,

[37] This formulation will have to be modified in the special case of statements explicitly about language, but I'll ignore this complication.

[38] Examples of each, in order: the phenomenalist Mill (1865), chs. 1 and 11, inspired by the idealist Berkeley (1710; 1713), (see the preface to Berkeley (1713)); the positivists, for example Ayer (1946, ch. 1); the idealist Brouwer (1913; 1949), see also Heyting (1931; 1966); the thoroughgoing verificationist Dummett (1975; 1977, ch. 7); and finally, Putnam (1977; 1980).

then, that there are, or at least may be, many truths about those entities that are beyond the reach of even our most idealized procedures of verification. Thus she could hardly be a verificationist.

So, even if holding a correspondence theory isn't the same thing as being a realist, it might seem that realism requires a correspondence theory. But this only follows on the assumption that these two candidates for a theory of truth—correspondence and verificationism—exhaust the field. They do not.

A third type of truth theory is based on the simple observation that 'so-and-so is true' says no more than 'so-and-so', in particular that 'My overcoat is in the closet' is true if and only if my overcoat is in the closet. This sort of theory has various names—redundancy theory, disappearance theory, deflationary theory—but I'll call it a disquotational theory. On this view, truth is a nothing more than a simplifying linguistic device. In cases like that of my overcoat, it does little more than stylistic work. When we use it in more complex contexts—for example, if I claim that everything in the Bible is true—it works as an abbreviating device that saves me a lot of time. With that one sentence I'm asserting that in the beginning God created the heavens and the earth, *and* now the earth was a formless void . . . and God's spirit hovered over the water, *and* God said, 'Let there be light', and there was light, *and* . . ., and so on through the many sentences in the Bible.

In recent years, there has been considerable debate between naturalized realists over correspondence versus disquotational truth.[39] The issue, of course, is whether or not the notion of correspondence truth must figure at all in our best theory of the world. For many purposes that seem to require a full-blown correspondence notion, disquotational truth has turned out to do the job as well. This is true even in philosophical contexts. For example, I claimed earlier that mathematical statements are true or false independently of our ability to know this. On the disquotational theory, this would come to a long (indeed infinite) conjunction: whether or not $2 + 2 = 4$ is independent of whether or not we can know which, *and* whether or not every non-empty bounded set of reals has a least upper bound is independent of whether not we can know which, *and* whether or not there is an inaccessible cardinal is independent of whether or not we can know

[39] See e.g. Field (1972), Grover *et al.* (1975), Leeds (1978), Devitt (1984), Field (1986), and the references cited there.

which, *and* . . ., etc. So the question is whether there are any jobs for which a notion of correspondence truth is actually indispensable.

To see what hangs on this, consider: Alfred Tarski's celebrated definition of correspondence truth reduces the problem of providing an account of that notion to the problem of providing an account of the word–world connections, that is, an account of the relation of reference that holds between a name and its bearer, between a predicate and the objects that satisfy it.[40] The disquotational theory, on the other hand, includes a theory of reference as unexciting as its theory of truth: 'Albert Einstein' refers to Albert Einstein; 'gold' refers to gold. So what hangs on the debate between correspondence and disquotational truth is the need for a substantive theory of what it is by virtue of which my use of the name 'Albert Einstein' manages to pick out that certain historical individual. This is no trivial matter; hence the lively debate over whether or not a correspondence theory is really necessary.

I won't get into that debate here, because it would take us too far afield and I certainly have nothing helpful to add, but I do want to take note of where the hunt has led these investigators. It's perhaps not surprising that where push seems to come to shove on the question of truth and reference is in that portion of our theory of the world that treats the activities of human beings.

Consider this case: Dr Jobe heals Isiah Thomas's ankle injury in time for the big game. How do we explain this phenomenon?[41]

As a first step, we notice that Dr Jobe has a vast number of true beliefs about sports injuries, about the rigours of basketball, about the physical and mental condition of basketball players, and about Isiah in particular. What Dr Jobe thinks about these things is usually correct. How, then, are we to explain his reliability on these topics? In an effort to answer this question, we begin to detail such things as the doctor's previous experience, his medical training, his interactions with many basketball players, and his previous interactions with Isiah himself.

Now the truth of the doctor's beliefs might be accounted for disquotationally. But the correspondence theorist will point out that among the sorts of connections between Jobe's beliefs and the subject matter of those beliefs that we've been describing in order to account for his reliability are just the sorts of connections that

[40] For Tarski's theory, see Tarski (1933); for the reduction, see Field (1972).
[41] Here I follow Field (1986), though the particular example is my own.

might well set up a robust referential connection between his uses of the predicate 'basketball player' and basketball players, between his use of 'Isiah Thomas' and the Piston guard.[42] So the debate between correspondence and disquotation comes down to the question of whether or not such a theory of reference can or need be constructed from these materials.

But notice, both parties to the debate agree that Jobe's reliability needs explanation, and they agree on the sorts of facts that might provide one. When a person is reliable on some subject—whether it be Jobe on Isiah's ankle, or a geologist on Mount St Helens, or a historian on the causes of the Industrial Revolution, or a ten-year-old kid on his favourite rock star—that reliability needs an explanation. We look for an account of how the person's cognitive machinery is connected back to what she's reliable about, via what she's read and the sources of that material, via her conversations with others and their sources, via her observations of indicators or instruments, and via her actual experience with her subject matter. The disagreement is only about whether or not this welter of material will produce a non-trivial theory of reference.

As I've said, I have no intention of taking sides on this last question; I'm perfectly willing to let the participants reach their own conclusions. The point that needs making here is this: even if the disquotationalist succeeds in relieving the realist of her dependence on correspondence truth, and thus on non-trivial reference, the matter of explaining the various 'reliable connections' will remain. Thus, in cases where the need for a referential connection seems to involve the mathematical realist in difficulties, casting off the need for reference is not likely to help, because the requirements of 'reliable connection' are almost certain to lead to the same, or essentially similar, difficulties.[43]

In what follows, then, I will sometimes phrase both the challenges to mathematical realism and my responses in terms of

[42] What the correspondence theorist has in mind here is a version of the causal theory of reference described in ch. 2, sect. 1, below.

[43] As Field (1986) makes clear, from the point of view of the overall realist project, a theory of reliable connection would probably be easier than one of full reference; for example, reference is compositional—the reference of a whole is thought to depend systematically on the reference of the parts—while a theory of reliability might not require so much detailed structure. But the aspects of the theory of reference that are thought to create difficulties for the mathematical realist are those it seems to share with the theory of reliability, for example, concerns about causation. See ch. 2, sect. 1, below.

correspondence truth and reference, but here, again for the record, I want to emphasize that this way of stating things is convenient but not essential. Those realists who believe that a robust reference relation is not needed in science, either because the disquotationalist is right, or for some other reason, are invited simply to recast the discussions that follow in terms of 'reliable connection'.

One final remark on realism and truth. Some anti-realists, assuming the realist is wedded to correspondence truth, have argued that realism is unscientific because it requires a connection between scientific theory and the world that reaches beyond the bounds of science itself.[44] Here the anti-realist attempts to saddle the realist with the now-familiar unnaturalized standpoint, the point of view that stands above, outside, or prior to, our best theories of the world, and from which is posed the question: what connects our theories to the world?

We've seen that in epistemology, the contemporary realist has answered by rejecting the extra-scientific challenge itself, along with the radical scepticism it engenders. The same goes for semantics. There is no point of view prior to or superior to that of natural science. What we want is a theory of how our language works, a theory that will become a chapter of that very scientific world-view. In order to arrive at this new chapter, it would be madness to cast off the scientific knowledge collected so far. Rather, we stand within our current best theory—what better account do we have of the way the world is?—and ask for an account of how our beliefs and our language connect up with the world as that theory says it is. This may be the robust theory of reference required by correspondence truth. If the disquotationalist is right, it may be something less structured, an account of reliable connection. But neither way is it something extra-scientific.

4. Realism in mathematics

Let me turn at last to realism in the philosophy of mathematics proper. Most prominent in this context is a folkloric position called 'Platonism' by analogy with Plato's realism about universals. As is

[44] See e.g. Putnam (1977), 125. Or, from a different point of view, Burgess (forthcoming *a*).

common with such venerable terms, it is applied to views of very different sorts, most of them not particularly Platonic.[45] Here I will take it in a broad sense as simply synonymous with 'realism' as applied to the subject matter of mathematics: mathematics is the scientific study of objectively existing mathematical entities just as physics is the study of physical entities. The statements of mathematics are true or false depending on the properties of those entities, independent of our ability, or lack thereof, to determine which.

Traditionally, Platonism in the philosophy of mathematics has been taken to involve somewhat more than this. Following some of what Plato had to say about his Forms, many thinkers have characterized mathematical entities as abstract—outside of physical space, eternal and unchanging—and as existing necessarily—regardless of the details of the contingent make-up of the physical world. Knowledge of such entities is often thought to be a priori—sense experience can tell us how things are, not how they must be—and certain—as distinguished from fallible scientific knowledge. I will call this constellation of opinions 'traditional Platonism'.

Obviously, this uncompromising account of mathematical reality makes the question of how we humans come to know the requisite a priori certainties painfully acute. And the successful application of mathematics to the physical world produces another mystery: what do the inhabitants of the non-spatio-temporal mathematical realm have to do with the ordinary physical things of the world we live in? In his theory of Forms, Plato says that physical things 'participate' in the Forms, and he uses the fact of our knowledge of the latter, via a sort of non-sensory apprehension, to argue that the soul must pre-exist birth.[46] But our naturalized realist will hardly buy this package.

Given these difficulties with traditional Platonism, it's not surprising that various forms of mathematical anti-realism have been proposed. I'll pause to consider a sampling of these views before describing the two main schools of contemporary Platonism.

[45] For example, though the term 'Platonism' suggests a realism about universals, many Platonists regard mathematics as the science of peculiarly mathematical particulars: numbers, functions, sets, etc. An exception is the structuralist approach considered in ch. 5, sect. 3, below.

[46] See his *Phaedo* 72 D–77 A.

In the late 1600s, in response to a number of questions from physical science, Sir Isaac Newton and Gottfried Wilhelm von Leibniz simultaneously and independently invented the calculus. Though the scientist's problems were solved, the new mathematical methods were scandalously error-ridden and confused. Among the most vociferous and perceptive critics was the idealist Berkeley, an Anglican bishop who hoped to silence the atheists by showing their treasured scientific thinking to be even less clear than theology. The central point of contention was the notion of infinitesimals, infinitely small amounts still not equal to zero, which Berkeley ridiculed as 'the ghosts of departed quantities'.[47] Two centuries later, Bolzano, Cauchy, and Weierstrass had replaced these ghosts with the modern theory of limits.[48]

This account of limits required a foundation of its own, which Georg Cantor and Richard Dedekind provided in their theory of real numbers, but these in turn reintroduced the idea of the completed infinite into mathematics. No one had ever much liked the seemingly paradoxical idea that a proper part of an infinite thing could be in some sense as large as the whole—there are as many even natural numbers as there are even and odd, there are as many points on a one-inch line segment as on a two-inch line segment—but the infinite sets introduced by Cantor and others gave rise to outright contradictions, of which Bertrand Russell's is the most famous:[49] consider the set of all sets that are not members of themselves. It is self-membered if and only if it isn't. The opening decades of this century saw the development of three great schools of thought on the nature of mathematics, all of them designed to deal in one way or another with the problem of the infinite.

The first of these is intuitionism, which dealt with the infinite by rejecting it outright. The original version of this position, first proposed by L. E. J. Brouwer,[50] was analogous to Berkeleian

[47] See Berkeley (1734), subtitled 'A Discourse Addressed to an Infidel Mathematician. Wherein It is Examined Whether the Object, Principles, and Inferences of the Modern Analysis are More Distinctly Conceived, or More Evidently Deduced, than Religious Mysteries and Points of Faith'. The quotation is from p. 89.

[48] For a more detailed description of the developments sketched in this paragraph and the next, see Kline (1972), chs. 17, 40, 41, and 51, or Boyer (1949).

[49] The paradox most directly associated with Cantor's work is Burali-Forti's (1897). See Cantor's discussion (1899). Russell's primary target was Frege, as will be noted below.

[50] Brouwer (1913; 1949). Other, less opaque, expositions of this position are Heyting (1931; 1966) and Troelstra (1969).

idealism: it takes the objects of mathematics to be mental constructions rather than objective entities. The modern version, defended by Michael Dummett,[51] is a brand of verificationism: a mathematical statement is said to be true if and only if it has been constructively proved. Either way, a series of striking consequences follow: statements that haven't been proved or disproved are neither true nor false; completed infinite collections (like the set of natural numbers) are illegitimate; much of infinitary mathematics must either be rejected (higher set theory) or radically revised (real number theory and the calculus).

These forms of intuitionism face many difficulties—e.g. does each mathematician have a different mathematics depending on what she's mentally constructed? how can we verify even statements about large finite numbers? etc.—but its most serious drawback is that it would curtail mathematics itself. My own working assumption is that the philosopher's job is to give an account of mathematics as it is practised, not to recommend sweeping reform of the subject on philosophical grounds. The theory of the real numbers, for example, is a fundamental component of the calculus and higher analysis, and as such is far more firmly supported than any philosophical theory of mathematical existence or knowledge. To sacrifice the former to preserve the latter is just bad methodology.

A second anti-realist position is formalism, the popular school of double-think mentioned above. The earliest versions of the view that mathematics is a game with meaningless symbols played heavily on a simple analogy between mathematical symbols and chess pieces, between mathematics and chess, but even its advocates were uncomfortably aware of the stark disanalogies:[52]

> To be sure, there is an important difference between arithmetic and chess. The rules of chess are arbitrary, the system of rules for arithmetic is such that by means of simple axioms the numbers can be referred to perceptual manifolds and can thus make [an] important contribution to our knowledge of nature.

The Platonist Gottlob Frege launched a fierce assault on early formalism, from many directions simultaneously, but the most

[51] Dummett (1975; 1977).

[52] Frege cites this quotation from Thomae in his critique of formalism: Frege (1903), § 88.

penetrating arose from just this point. It isn't hard to see how various true statements of mathematics can help me determine how many bricks it will take to cover the back patio, but how can a meaningless string of symbols be any more relevant to the solution of real world problems than an arbitrary arrangement of chess pieces?

This is Frege's problem: what makes these meaningless strings of symbols useful in applications?[53] Suppose, for example, that a physicist tests a hypothesis by using mathematics to derive an observational prediction. If the mathematical premiss involved is just a meaningless string of symbols, what reason is there to take that observation to be a consequence of the hypothesis? And if it is not a consequence, it can hardly provide a fair test. In other words, if mathematics isn't true, we need an explanation of why it is all right to treat it as true when we use it in physical science.

The most famous version of formalism, the one expounded during the period under consideration here, was David Hilbert's programme.[54] Hilbert, like Brouwer, felt that only finitary mathematics was truly meaningful, but he considered Cantor's theory of sets 'one of the supreme achievements of purely intellectual human activity' and promised, in a famous remark, that

> No one shall drive us out of the paradise which Cantor has created for us. (Hilbert (1926), 188, 191)

Hilbert proposed to save infinitary mathematics by treating it instrumentally—meaningless statements about the infinite are a useful tool in deriving meaningful statements about the finite—but he, unlike the scientific instrumentalists, was sensitive to the question of how this practice could be justified. Hilbert's plan was to give a metamathematical proof that the use of the meaningless statements of infinitary mathematics to derive meaningful statements of finitary mathematics would never produce incorrect finitary results. The same line of thought might have applied to its use in natural science as well, thus solving Frege's problem. Hilbert's efforts to carry through on this project produced the rich

[53] See Frege (1903), § 91.
[54] See Hilbert (1926; 1928).

new field of metamathematics, but Kurt Gödel soon proved that its cherished goal could not be reached.[55]

For all the simplicity of game formalism and the fame of Hilbert's programme, many mathematicians, when they claim to be formalists, actually have another idea in mind: mathematics isn't a science with a peculiar subject matter; it is the logical study of what conclusions follow from which premisses. Philosophers call this position 'if-thenism'. Several prominent philosophers of mathematics have held this position at one time or another—Hilbert (before his programme), Russell (before his logicism), and Hilary Putnam (before his Platonism)[56]—but all ultimately rejected it. Let me briefly indicate why.

A number of annoying difficulties plague the if-thenist: which logical language is appropriate for the statement of premisses and conclusions? which premisses are to be presupposed in cases like number theory, where assumptions are usually left implicit? from among the vast range of arbitrary possibilities, why do mathematicians choose the particular axiom systems they do to study? what were historical mathematicians doing before their subjects were axiomatized? what are they doing when they propose new axioms? and so on. But the question that seems to have scotched if-thenism in the minds of Russell and Putnam was a version of Frege's problem: how can the fact that one mathematical statement follows from another be correctly used in our investigation of the physical world? The general thrust of the if-thenist's reply seems to be that the antecedent of a mathematical if-then statement is treated as an idealization of some physical statement. The scientist then draws as a conclusion the physical statement that is the unidealization of the consequent.[57]

Notice that on this picture, the physical statements must be entirely mathematics-free; the only mathematics involved is that used in moving between them. Unfortunately, many of the

[55] See Gödel (1931). Enderton (1972), ch. 3, gives a readable presentation. Detlefsen (1986) attempts to defend Hilbert's programme against the challenge of Gödel's theorem. Simpson (1988) and Feferman (1988) pursue partial or relativized versions within the limitations of Gödel's theorem.

[56] See Resnik (1980), ch. 3, for discussion. There if-thenism is called 'deductivism'. See also Putnam (1979), p. xiii. Russell's logicism and Putnam's Platonism will be considered below.

[57] See Korner (1960), ch. 8. Cf. Putnam (1967*b*), 33.

statements of physical science seem inextricably mathematical. To quote Putnam, after his conversion:

> one wants to say that the Law of Universal Gravitation makes an objective statement about bodies—not just about sense data or meter readings. What is the statement? It is just that bodies behave in such a way that the quotient of two numbers *associated* with the bodies is equal to a third number *associated* with the bodies. But how can such a statement have any objective content at all if numbers and 'associations' (i.e. functions) are alike mere fictions? It is like trying to maintain that God does not exist and angels do not exist while maintaining at the very same time that it is an objective fact that God has put an angel in charge of each star and the angels in charge of each of a pair of binary stars were always created at the same time! If talk of numbers and 'associations' between masses, etc. and numbers is 'theology' (in the pejorative sense), then the Law of Universal Gravitation is likewise theology. (Putnam (1975*b*), 74–5)

In other words, the if-thenist account of applied mathematics requires that natural science be wholly non-mathematical, but it seems unlikely that science can be so purified.[58]

The third and final anti-realist school of thought I want to consider here is logicism, or really, the version of logicism advanced by the logical positivists. Frege's original logicist programme aimed to show that arithmetic is reducible to pure logic, that is, that its objects—numbers—are logical objects and that its theorems can be proved by logic alone.[59] This version of logicism is outright Platonistic: arithmetic is the science of something objective (because logic is objective), that something objective consists of objects (numbers), and our logical knowledge is a priori. If this project had succeeded, the epistemological problems of Platonism would have been reduced to those of logic, presumably a gain. But Frege's project failed; his system was inconsistent.[60] Russell and Whitehead took up the banner in their *Principia Mathematica*, but were forced to adopt fundamental assumptions no one accepted as

[58] Hartry Field's ambitious attempt to do this will be considered in ch. 5, sect. 2, below. See Field (1980; 1989).

[59] See Frege (1884).

[60] The trouble was the original version of Russell's paradox. (See Russell's letter to Frege, Russell (1902).) Frege's numbers were extensions of concepts. (See ch. 3 below.) Some concepts, like 'red', don't apply to their extensions, others, like 'infinite', do. Russell considered the extension of the concept 'doesn't apply to its own extension'. If it applies to its own extension then it doesn't, and vice versa. This contradiction was provable from Frege's fundamental assumptions. There have been efforts to revive Frege's system; see e.g. Wright (1983) and Hodes (1984).

purely logical.[61] Eventually, Ernst Zermelo (aided by Mirimanoff, Fraenkel, Skolem, and von Neumann) produced an axiom system that showed how mathematics could be reduced to set theory,[62] but again, no one supposed that set theory enjoys the epistemological transparency of pure logic.

Still, the idea that mathematics is just logic was not dead; it was taken up by the positivists, especially Rudolf Carnap.[63] For these thinkers, however, there are no logical objects of any kind, and the laws of logic and mathematics are true only by arbitrary convention. Thus mathematics is not, as the Platonist insists, an objective science. The advantage of this counterintuitive view is that mathematical knowledge is easily explicable; it arises from human decisions. Question: Why are the axioms of Zermelo–Fraenkel true? Answer: Because they are part of the language we've adopted for using the word 'set'.

This conventionalist line of thought was subjected to a historic series of objections by Carnap's student, W. V. O. Quine.[64] The key difficulty is that both mathematical and physical assumptions are enshrined in Carnap's official language. How are we to separate the conventionally adopted mathematical part of the language from the factually true physical hypotheses? Quine argues that it isn't enough to say that the scientific claims, not the mathematical ones, are supported by empirical data:

> The semblance of a difference in this respect is largely due to overemphasis of departmental boundaries. For a self-contained theory which we can check with experience includes, in point of fact, not only its various theoretical hypotheses of so-called natural science but also such portions of logic and mathematics as it makes use of. (Quine (1954), 367)

Mathematics is part of the theory we test against experience, and a successful test supports the mathematics as much as the science.

Carnap makes several efforts to separate mathematics from natural science, culminating in his distinction between analytic and synthetic. Mathematical statements, he argues, are analytic, that is,

[61] See Russell and Whitehead (1913).

[62] Zermelo's first presentation is Zermelo (1908*b*). See also Mirimanoff (1917*a*, *b*), Fraenkel (1922), Skolem (1923), and von Neumann (1925). The standard axioms are now called 'Zermelo–Fraenkel set theory' or ZFC (ZF when the axiom of choice is omitted). See Enderton (1977), 271–2.

[63] See Carnap (1937; 1950).

[64] See Quine (1936; 1951; 1954).

true by virtue of the meanings of the words involved (the logical and mathematical vocabulary); scientific statements, on the other hand, are synthetic, true by virtue of the way the world is. Quine examines this distinction in great detail, investigating various attempts at clear formulation, and concludes:

It is obvious that truth in general depends on both language and extralinguistic fact. The statement 'Brutus killed Caesar' would be false if the world had been different in certain ways, but it would also be false if the word 'killed' happened rather to have the sense of 'begat'. Thus one is tempted to suppose in general that the truth of a statement is somehow analyzable into a linguistic component and a factual component. Given this supposition, it next seems reasonable that in some statements the factual component should be null; and these are the analytic statements. But, for all its a priori reasonableness, a boundary between analytic and synthetic statements simply has not been drawn. That there is such a distinction to be drawn at all is an unempirical dogma of empiricists, a metaphysical article of faith. (Quine (1951), 36–7)

Without a clear distinction between analytic and synthetic, Carnap's anti-Platonist version of logicism fails.

I will leave the three great schools at this point. I don't claim to have refuted either formalism or conventionalism, though I hope the profound difficulties they face have been drawn clearly enough. Intuitionism I reject on the grounds given above; I assume that the job of the philosopher of mathematics is to describe and explain mathematics, not to reform it.

Let me return now to Platonism, the view that mathematics is an objective science. Platonism naturally conflicts with each of the particular forms of anti-realism touched on here—with intuitionism on the objectivity of mathematical entities, with formalism on the status of infinitary mathematics, with logicism on the need for mathematical existence assumptions going beyond those of logic—but the Platonist's traditional and purest opponent is the nominalist, who simply holds that there are no mathematical entities. (The term 'nominalism' has followed 'Platonism' in its migration from the debate over universals into the debate over mathematical entities.) Two forms of Platonism dominate contemporary debate. The first of these derives from the work of Quine and Putnam sketched above—their respective criticisms of conventionalism and if-thenism—and the second is described by Gödel as the philosophical underpinning for his famous theorems.[65] As Quine and Putnam's writings have just been discussed, let me begin with them.

Quine's defence of mathematical realism follows directly on the heels of the defences of common-sense and scientific realism sketched above. On the naturalized approach, we judge what entities there are by seeing what entities we need to produce the most effective theory of the world. So far, these include medium-sized physical objects and the theoretical entities of physical science, and so far, the nominalist might well agree. But if we pursue the question of mathematical ontology in the same spirit, the nominalist seems cornered:

> A platonistic ontology . . . is, from the point of view of a strictly physicalistic conceptual scheme, as much a myth as that physicalistic conceptual scheme itself is for phenomenalism. This higher myth is a good and useful one, in turn, in so far as it simplifies our account of physics. Since mathematics is an integral part of this higher myth, the utility of this myth for physical science is evident enough. (Quine (1948), 18)

If we countenance an ontology of physical objects and unobservables as part of our best theory of the world, how are we to avoid countenancing mathematical entities on the same grounds? Carnap suggested what Quine calls a 'double standard'[66] in ontology, according to which questions of mathematical existence are linguistic and conventional and questions of physical existence are scientific and real, but we've already seen that this effort fails.

We've also seen that Putnam takes the same thinking somewhat further, emphasizing not only that mathematics simplifies physics, but that physics can't even be formulated without mathematics:[67] 'mathematics and physics are integrated in such a way that it is not possible to be a realist with respect to physical theory and a nominalist with respect to mathematical theory' (Putnam (1975*b*), 74). He concludes that talk about[68]

> mathematical entities is indispensable for science . . . therefore we should

[65] See his letters to Wang, quoted in Wang (1974*b*), 8–11, and Feferman's discussion (1984*b*).

[66] Quine (1951), 45.

[67] See the long quotation from Putnam (1975*b*) above. A more complete account appears in Putnam (1971), esp.§§ 5 and 7.

[68] He really says 'quantification over', which derives from Quine's official criterion of ontological commitment (1948), but I don't want to get into the debate over that precise formulation.

accept such [talk]; but this commits us to accepting the existence of the mathematical entities in question. This type of argument stems, of course, from Quine, who has for years stressed both the indispensability of [talk about] mathematical entities and the intellectual dishonesty of denying the existence of what one daily presupposes. (Putnam (1971), 347)

We are committed to the existence of mathematical objects because they are indispensable to our best theory of the world and we accept that theory.

The particular brand of Platonism that arises from these Quine/Putnam indispensability arguments has some revolutionary features. Recall that traditional Platonism takes mathematical knowledge to be a priori, certain, and necessary. But, if our knowledge of mathematical entities is justified by the role it plays in our empirically supported scientific theory, that knowledge can hardly be classified as a priori.[69] Furthermore, if we prefer to alter our scientific hypotheses rather than our mathematical ones when our overall theory meets with disconfirmation, it is only because the former can usually be adjusted with less perturbation to the theory as a whole.[70] Indeed, Putnam[71] goes so far as to suggest that the best solution to difficulties in quantum mechanics may well be to alter our logical laws rather than any physical hypotheses. Thus the position of mathematics as part of our best theory of the world leaves it as liable to revision as any other part of that theory, at least in principle, so mathematical knowledge is not certain. Finally, the case of necessity is less clear, if only because Quine rejects such modal notions out of hand, but the fact that our mathematics is empirically confirmed in this world surely provides little support for the claim that it is likely to be true in some other possible circumstance. So Quine/Putnam Platonism stands at some considerable remove from the traditional variety.

But while disagreement with a venerable philosophical theory is no clear demerit, disagreement with the realities of mathematical practice is. First, notice that unapplied mathematics is completely without justification on the Quine/Putnam model; it plays no indispensable role in our best theory, so it need not be accepted:[72]

[69] See Putnam (1975*b*) for an explicit discussion of a posteriori methods in mathematics. Kitcher (1983) attacks the idea that mathematics is a priori from a different angle.

[70] See Quine (1951), 43–4.

[71] Putnam (1968).

[72] See also Putnam (1971), 346–7.

> So much of mathematics as is wanted for use in empirical science is for me on a par with the rest of science. Transfinite ramifications are on the same footing insofar as they come of a simplificatory rounding out, but anything further is on a par rather with uninterpreted systems. (Quine (1984), 788)

Now mathematicians are not apt to think that the justification for their claims waits on the activities in the physics labs. Rather, mathematicians have a whole range of justificatory practices of their own, ranging from proofs and intuitive evidence, to plausibility arguments and defences in terms of consequences. From the perspective of a pure indispensability defence, this is all just so much talk; what matters is the application.

If this weren't enough to disqualify Quine/Putnamism as an account of mathematics as it is practised, consider one last point. In this picture of our scientific theorizing, mathematics enters only at fairly theoretical levels. The most basic evidence takes the form of non-mathematical observation sentences—e.g. 'this chunk of gold is malleable'—and the initial levels of theory consist of non-mathematical generalizations—'gold is a malleable metal'. Mathematics only enters the picture at the more theoretical levels—'gold has atomic number 79'—so it is on an epistemic par with this higher-level theory.[73] But isn't it odd to think of '2 + 2 = 4' or 'the union of the set of even numbers with the set of odd numbers is the set of all numbers' as highly theoretical principles? In Charles Parsons's phrase, Quine/Putnamism 'leaves unaccounted for precisely the *obviousness* of elementary mathematics'.[74]

By way of contrast, the Gödelian brand of Platonism takes its lead from the actual experience of doing mathematics, which he takes to support Platonism as suggested in section 1 above. For Gödel, the most elementary axioms of set theory *are* obvious; in his words, they 'force themselves upon us as being true'.[75] He accounts for this by positing a faculty of mathematical intuition that plays a role in mathematics analogous to that of sense perception in the physical sciences, so presumably the axioms force themselves upon us as explanations of the intuitive data much as the assumption of medium-sized physical objects forces itself upon us as an explanation of our sensory experiences. To push this analogy, recall that this style of argument for common-sense realism might have been

[73] See Quine (1948), 18–19.
[74] Parsons (1979/80), 151. See also Parsons (1983*b*).
[75] Gödel (1947/64), 484.

undercut if phenomenalists had succeeded in giving non-realistic translations of our physical object statements. Similarly, Gödel notes that Russell's 'no-class' interpretation of *Principia* was an effort to do the work of set theory, that is, to systematize all of mathematics, without sets. Echoing the common-sense realist, Gödel takes the failure of Russell's project as support for his mathematical realism:

This whole scheme of the no-class theory is of great interest as one of the few examples, carried out in detail, of the tendency to eliminate assumptions about the existence of objects outside the 'data' and to replace them by constructions on the basis of these data.[76] The result has been in this case essentially negative . . . All this is only a verification of the view defended above that logic and mathematics (just as physics) are built up on axioms with a real content which cannot be 'explained away'. (Gödel (1944), 460–1)

He concludes that

the assumption of [sets] is quite as legitimate as the assumption of physical bodies and there is quite as much reason to believe in their existence. They are in the same sense necessary to obtain a satisfactory system of mathematics as physical bodies are necessary for a satisfactory theory of our sense perceptions . . . (Gödel (1944), 456–7)

But this analogy of intuition with perception, of mathematical realism with common-sense realism, is not the end of Gödel's elaboration of the mathematical realist's analogy between mathematics and natural science. Just as there are facts about physical objects that aren't perceivable, there are facts about mathematical objects that aren't intuitable. In both cases, our belief in such 'unobservable' facts is justified by their role in our theory, by their explanatory power, their predictive success, their fruitful interconnections with other well-confirmed theories, and so on. In Gödel's words:

even disregarding the [intuitiveness] of some new axiom, and even in case it has no [intuitiveness] at all, a probable decision about its truth is possible also in another way, namely, inductively by studying its 'success'. . . . There might exist axioms so abundant in their verifiable consequences,

[76] In this passage, 'data' means 'logic without the assumption of the existence of classes' (Gödel (1944), 460 n. 22). Earlier in this same paper, Gödel refers to arithmetic as 'the domain of the kind of elementary indisputable evidence that may be most fittingly compared with sense perception' (p. 449).

shedding so much light upon a whole field, and yielding such powerful methods for solving problems . . . that, no matter whether or not they are [intuitive], they would have to be accepted at least in the same sense as any well-established physical theory. (Gödel (1947/64), 477)

Quite a number of historical and contemporary justifications for set theoretic hypotheses take this form, as will come out in Chapter 4. Here the higher, less intuitive, levels are justified by their consequences at lower, more intuitive, levels, just as physical unobservables are justified by their ability to systematize our experience of observables. At its more theoretical reaches, then, Gödel's mathematical realism is analogous to scientific realism.

Thus Gödel's Platonistic epistemology is two-tiered: the simpler concepts and axioms are justified intrinsically by their intuitiveness; more theoretical hypotheses are justified extrinsically, by their consequences. This second tier leads to departures from traditional Platonism similar to Quine/Putnam's. Extrinsically justified hypotheses are not certain,[77] and, given that Gödel allows for justification by fruitfulness in physics as well as in mathematics,[78] they are not a priori either. But, in contrast with Quine/Putnam, Gödel gives full credit to purely mathematical forms of justification —intuitive self-evidence, proofs, and extrinsic justifications within mathematics—and the faculty of intuition does justice to the obviousness of elementary mathematics.

Among Gödel's staunchest critics is Charles Chihara.[79] Even if Gödel has succeeded in showing that the case for the existence of mathematical entities runs parallel to the case for the existence of physical ones, Chihara argues that he has by no means shown that the two cases are of the same strength, and thus, that he has not established that there is as much reason to believe in the one as to believe in the other.[80] Furthermore, Chihara argues, the existence of mathematical entities is not required to explain the experience of mathematical intuition and agreement:

I believe it is at least as promising to look for a naturalistic explanation based on the operations and structure of the internal systems of human beings. (Chihara (1982), 218)

[77] Gödel (1944), 449.
[78] Gödel (1947/64), 485.
[79] See Chihara (1973), ch. 2; (1982).
[80] Chihara (1982), 213–14.

. . . mathematicians, regarded as biological organisms, are basically quite similar. (Chihara (1973), 80)

And finally, he questions whether Gödel's intuition offers any explanation at all:[81]

the 'explanation' offered is so vague and imprecise as to be practically worthless: all we are told about how the 'external objects' explain the phenomena is that mathematicians are 'in some kind of contact' with these objects. What empirical scientist would be impressed by an explanation this flabby? (Chihara (1982), 217)

Now the Gödelian Platonist is not entirely defenceless in the face of this attack. For example, Mark Steiner[82] points out that Chihara's 'explanation' is likewise lacking in muscle tone: the similarity of human beings as organisms can hardly explain their agreement about mathematics when it is consistent with so much disagreement on other subjects. Still, most observers tend to agree that no appeal to purported human experiences of *x*s that underlie our theory of *x*s can justify a belief in the existence of *x*s unless we have some independent reason to think our theory of *x*s is true.[83] Thus the purported human dealings with witches that underlie our theory of witches don't justify a belief in witches unless we have some independent reason to think that our theory of witches is actually correct.

But notice: we have recently rehearsed just such an independent reason in the case of mathematics, namely, the indispensability arguments of Quine and Putnam. Unless endorsing these commits one to the view that there is no peculiarly mathematical form of evidence—and I don't see why it should[84]—there is room for an attractive compromise between Quine/Putnam and Gödelian Platonism. It goes like this: successful applications of mathematics give us reason to believe that mathematics is a science, that much of it at least approximates truth. Thus successful applications justify, in a general way, the practice of mathematics. But, as we've seen, this isn't enough to give an adequate account of mathematical practice,

[81] These remarks of Chihara's are actually addressed to a quotation from Kreisel, but it is clear from the context that he thinks the same objection applies to Gödel's intuition.

[82] Steiner (1975*b*), 190.

[83] See Steiner (1975*b*), 190. For a similar sentiment, see Putnam (1975*b*), 73–4.

[84] Nor does Parsons (1983*b*), 192–3.

of how and why it works. We still owe an account of the obviousness of elementary mathematics, which Gödel's intuition is designed to provide, and an account of other purely mathematical forms of evidence, like proof and various extrinsic methods. This means we need to explain what intuition is and how it works; we need to catalogue extrinsic methods and explain why they are rational methods in the pursuit of truth.

From Quine/Putnam, this compromise takes the centrality of the indispensability arguments; from Gödel, it takes the recognition of purely mathematical forms of evidence and the responsibility for explaining them. Thus it averts a major difficulty with Quine/Putnamism—its unfaithfulness to mathematical practice—and a major difficulty with Gödelism—its lack of a straightforward argument for the truth of mathematics. But whatever its merits, compromise Platonism does nothing to remedy the flabbiness of Gödel's account of intuition. And it is in this neighbourhood that many contemporary objections to Platonism are concentrated.[85]

I opened this chapter with the hope of reinstating the mathematician's pre-philosophical realism, of devising a defensible refinement of that attitude that remains true to the phenomenology of practice. Along the way, I've sided with common-sense realism, scientific realism, and philosophical naturalism, and seconded many of the advances of Quine/Putnam and Gödelian Platonism. It will come as no surprise, then, that the position to be defended here is a version of compromise Platonism. I'll call it 'set theoretic realism'.

Chapter 2 outlines a naturalistic epistemology for items located on the lower tier of Gödel's two-tiered epistemology, a replacement for Gödel's intuition. The ontological question of the relationship between sets and other mathematical entities, particularly natural and real numbers, is the subject of Chapter 3. Chapter 4 contains some preliminary spadework on the problem of theoretical justification, the second of Gödel's two tiers. I argue that this ill-understood problem is the most important open question of our day, not only for set theoretic realism, but for many other mathematical philosophies as well. Chapter 5 takes a final look at set theoretic realism from physicalist and structuralist perspectives.

[85] See ch. 2, sect. 1, below.

2

PERCEPTION AND INTUITION

1. What is the question?

The general outlines of the epistemological challenge to Platonism have already been hinted at, but I'd like now to place the problem in the context of contemporary philosophy. The sense that there is a problem goes back, as we've seen, to Plato himself, but the modern form, the one exhaustively discussed in the contemporary literature, derives from Paul Benacerraf's 'Mathematical truth', which appeared in the early seventies.[1] Since then, it has become commonplace for scholarly writings on the philosophy of mathematics to begin by dismissing Platonism on the basis of Benacerraf's argument. Benacerraf himself draws no such dogmatic conclusion, but his successors, even those with generally realistic leanings, have scorned Platonism.[2]

The Benacerrafian syllogism rests on two premisses. The second is a traditional Platonistic account of the nature of mathematical entities as abstract, in particular, as non-spatio-temporal. The first premiss concerns the nature of human knowledge: what is it for me to know something? It was originally suggested, again by Plato,[3] that it is enough that I believe it, that my belief be justified, and that the belief be true. Though Plato raised some objections of his own to this 'justified, true belief' account of knowledge, it wasn't until 1963 that Edmund Gettier pointed out what is now considered its fatal weakness.[4]

Suppose I see Dick driving a Hillman; suppose he offers me a ride

[1] Benacerraf (1973).

[2] The anti-Platonisms of Field (1980), Bonevac (1982), Gottlieb (1980), and Hellman (1989), are all at least partly motivated by Benacerrafian considerations. This style of argument is also noted with approval by Kitcher (1983), 59, and Resnik (1981), 529, (1982), 95, and (forthcoming *a*, *b*).

[3] See his *Theaetetus*, 202 C.

[4] Gettier (1963).

to work in this car. On the basis of this experience, I come to believe that Dick owns a Hillman. My belief that Dick owns a Hillman is surely justified—he gave me a lift in one—and let us further suppose that it is true—that Dick does indeed own a Hillman. But—and here's the catch—he doesn't own *this* Hillman. The Hillman Dick actually owns is in the shop, as it often is, and this one, the one I saw, the one I rode in, was borrowed from Frank. In this case, though I have a justified, true belief that he does, I can't be said to *know* Dick owns a Hillman. For knowledge, there is some further requirement.

Some years after Gettier's paper came a response from Alvin Goldman,[5] diagnosing the problem in cases like mine and Dick's, and proposing a fourth clause in the definition of knowledge to cure it. The difficulty, according to Goldman and many others who largely agreed with him,[6] is that Dick's Hillman was not the car that caused me to believe as I did. For a justified, true belief to count as knowledge, what makes the belief true must be appropriately[7] causally responsible for that belief. This idea, in its many versions, is called the 'causal theory of knowledge'.

The two premisses, then, of our Benacerrafian argument are the causal theory of knowledge and the abstractness of mathematical objects. What makes '2 + 2 = 4' true is the nature of the abstract entities 2 and 4 and the operation plus; for me to know that '2 + 2 = 4', those entities must play an appropriate causal role in the generation of my belief. But how can entities that don't even inhabit the physical universe take part in any causal interaction whatsoever? Surely to be abstract is also to be causally inert. Thus, if Platonism is true, we can have no mathematical knowledge. Assuming that we do have such knowledge, Platonism must be false.

This dramatic conclusion can be pushed further by recent progress in the theory of reference.[8] How does a name pick out a thing? In this field, the classical theory is Frege's:[9] a name is associated with a description that uniquely identifies the thing the name names; for example, 'Isiah Thomas' is associated with the description 'best friend of Magic Johnson'. Numerous variations on

[5] Goldman (1967).

[6] See e.g. Skyrms (1967) or Harman (1973).

[7] Suppose, by some neural fluke, my justified true belief is caused by my being hit on the head by Dick's Hillman. This would be inappropriate.

[8] See Lear (1977).

[9] See Frege (1892*b*).

this idea have been proposed—that there are in fact many descriptions associated with the name, that some of these might even be false, that what counts is the truth of a sufficient number of the more important ones, and so on[10]—but the central descriptive character of the referring relation remains.

In 1972, Saul Kripke[11] called this account into question. Suppose I hear the name 'Einstein' used repeatedly in discussions to which I am not very attentive, and I come to believe only one thing about him, namely, that he invented the atom bomb. Of course, Einstein didn't invent the atom bomb, but this is nevertheless the one and only description I associate with the name. On the description theory, my use of the name should refer to someone else, or to no one, if no single person invented the bomb. That this isn't the case is made clear by the reaction of my physicist friend who insists that I'm dead wrong in my belief about Einstein. If the description theory were correct, I'd have made a true statement about someone else—the person who did invent the bomb—or a truth-valueless statement about nobody—if no one person invented the bomb—but in fact I made a false statement about Einstein himself.

Kripke and others[12] react to this problem by proposing a very different picture of how we refer. My use of the name 'Einstein' picks out Einstein, not by virtue of my knowledge of some uniquely identifying description of the man, but because my usage is borrowed from those I heard using it, theirs in turn borrowed from their teachers or from books, the usage there borrowed from someone else's, in a chain leading back, ultimately, to someone who was in a position to dub the actual individual. Thus my use of 'Einstein' refers to Einstein, despite my ignorance, because it is part of a network of borrowed usage that extends from me back to a somewhat imaginary event called an 'initial baptism' in which Einstein himself participated.

A similar story works for scientific general terms like 'gold': a chain of communication leads back to an event in which the baptist isolated some samples of the metal and declared that 'this and stuff like it is gold'. Thus the scientific community was able to refer to

[10] See e.g. Searle (1958), Strawson (1959), ch. 6.

[11] Kripke (1972). For further discussion of the following and other examples, see Devitt (1981), 13–20, and Salmon (1981), 23–32.

[12] Most notably, Putnam (1975*a*), chs. 11–13. See also Devitt (1981) and references cited there.

gold via a direct connection with samples even before it knew enough about atomic weights to give a uniquely identifying description. And successive, very different scientific theories can be about the same things, because the referents are picked out by chains leading back to the dubbing of samples, not by the very different, often erroneous, descriptions the competing theories espouse.

Of course not all general terms fit this picture. 'Bachelor', for example, refers to whatever satisfies the description 'unmarried male', not to things more or less like Marcel Proust in some yet-to-be-discovered respect. The theory works, not when we have an explicit description or definition in mind for which our term is an abbreviation, but when we notice a similarity between various things, dub these and things like them by some term, and then set out to discover the underlying traits that make these things what they are. Such groupings, common in science, are called 'natural', as opposed to 'nominal', kinds.[13]

The kind consisting of all mathematical objects seems unlikely to be nominal, because available descriptions tend to be blatantly circular ones like 'what mathematicians study'. Rather, in picking out the kind, we get our point across by examples: mathematical objects are numbers, sets, functions, Hilbert spaces, and things like that. But, some might argue, if all mathematics is reducible to set theory, or if we simply restrict our attention to set theory, there is a simple definition after all, namely, that a set is a thing that occurs in the iterative hierarchy.[14]

Two things scotch this suggestion. First, the definition in terms of the iterative conception is still circular; we have to know what a set is, indeed what an arbitrary subset is, before we can understand it. Considering this problem, I suppose there's little need for a second objection, but I want to point out that the iterative conception is a

[13] These natural kinds are the natural collections mentioned in connection with universals in ch. 1, sect 2, above. See Quine (1969*d*), and Ayers (1981) for very different discussions of these ideas.

[14] The iterative hierarchy is arranged in stages. The first stage consists of whatever individuals we begin with. (In pure set theory, this is the empty set.) The second stage consists of the subsets of the first; the third of the subsets of the first and second; and so on. The first infinite stage, stage ω, consists of all the sets generated at the finite levels. Stage $\omega + 1$ includes all subsets of stage ω. And so on. Enderton (1977), 7–9, gives an informal introduction; for a more complete discussion, see Boolos (1971) or Shoenfield (1977).

theory of what sets are in much the same sense as 'having atomic number 79' is a theory of what gold is. Sets of natural numbers and point sets were considered long before Zermelo proposed his 1908 axiomatization, and the iterative picture was first described only years after that, in 1930.[15] In both cases, we start with samples, dub the kind, then go on to investigate the nature of that kind; in both the cases, the natural kind as identified by samples pre-dated the scientific theory and would survive its demise. If our theory of the atomic structure of gold turned out to be incorrect, we could go on to form another theory of the same stuff; similarly, theoretical considerations might lead us to drop the claim that all sets occur in the iterative hierarchy and to study non-well-founded sets as well.[16]

If the most basic mathematical kinds are natural rather than nominal, the referential difficulty for Platonism arises when we consider the nature of the required initial baptism. In the simplest case, we imagine the baptist standing in front of a number of samples of the natural kind in question and declaring that 'these and things like them are . . .'. In the case of gold, there is a direct causal interaction with the samples, that is, they are touched, seen, perceived. Because the links in the chain of communication are causal, and the direct connection between the baptist and the samples is causal,[17] this theory is often called 'the causal theory of reference'. Running parallel to Benacerraf's epistemological dilemma, we have two premisses—the causal theory of reference and the abstractness, and hence causal inertness, of mathematical objects—that lead to another unpalatable conclusion for the Platonist: we can't refer to mathematical objects. If mathematical reality is as the Platonist says it is, we are doomed not only to

[15] See Zermelo (1908*b*) for the first explicit statement of set theoretic axioms. The iterative conception is described in Zermelo (1930), though some (see Wang (1974*a*)) give prior credit to Mirimanoff (1917*a*, *b*).

[16] Indeed, the axiom of foundation, which restricts the range of set theoretic study to the members of the iterative hierarchy, is often taken not as a truth, but as a simplifying assumption. See Maddy (1988*a*), § 1.2. On the possibility of non-well-founded sets, see Aczel (1988).

[17] Kripke allows for baptism by description as well as by ostension—for example, 'Gold is the stuff found in Fort Knox'—but an attempted descriptive baptism of sets would necessarily run in a circle—'Sets are things like the set of my two hands'. Besides which, Devitt has argued (1981, pp. 36–42) that when a description is used referentially rather than attributively—as it is in a descriptive baptism—a causal grounding is still required. (Kripke's own attitude is not altogether clear. See Kim (1977), 615–17.)

ignorance, but to silence as well. And again, assuming that we can talk about mathematical objects, Platonism must be false.

These, then, are the serious epistemological challenges many philosophers take to have sunk the Platonist's ship. Some have replied that the causal theories are irrelevant to mathematics, because they are theories of a posteriori, contingent knowledge and mathematical knowledge is a priori and necessary, but this sort of response is of no use to the compromise Platonist who follows Quine in questioning these distinctions.[18] Let me take a moment now to examine the cogency of the causal arguments from this perspective.

We are faced with two anti-Platonist arguments. These arguments seem to depend on causal theories of knowledge and reference, along with the traditional Platonistic account of the nature of mathematical entities. Thus one pro-Platonist approach would be to call the causal theories into question. In fact, this approach was adopted early on, by Mark Steiner, in a paper that appeared in the same year as Benacerraf's.[19]

Steiner's argument is in two parts. First, he finds fault with several particular formulations of the causal theory and concludes that even its best formulation is implausible. He then argues that a suitably generalized 'causal theory' allows that a fact about numbers, for example, might play a role in the causal explanation of my belief in the corresponding axiom of number theory after all, simply because the axioms of number theory and analysis will all figure in any such explanation.[20] He summarizes: 'the most plausible version of the causal theory of knowledge admits Platonism, and the version most antagonistic to Platonism is implausible' (Steiner (1975*a*), 116).

Now there is room for rebuttal to both parts of Steiner's claim. For example, the case Steiner uses to sink the 'best' version of the

[18] See Wright (1983), § xi, and Hale (1987), 86–90, for the a priorist line of thought, and Lewis (1986), 108–15, for the necessitarian. It might also be argued that the causal condition isn't required for mathematical knowledge because there is no such thing as a mathematical Gettier case, that is, a mathematical case in which the subject has a justified true belief without knowledge. But such cases can arise, if not for all compromise Platonists, at least for the set theoretic realist. See sect. 2 below.

[19] See Steiner (1973), which later became ch. 4 of his (1975*a*).

[20] This thinking rests on the Quinean idea that mathematics will be part of the overall theory that causally explains my belief in the axioms of number theory.

causal theory involves inferential knowledge, but Benacerraf's version of the causal theory doesn't require that inferential knowledge meet the causal condition placed on more basic, non-inferential knowledge.[21] On the other hand, again for example, there is no guarantee that the particular axiom of number theory I believe will in fact figure in the causal explanation of that belief,[22] and even if it did, would it figure 'appropriately'? But most interesting for our assessment of the problem is the strong reaction of one reviewer:

> it is a crime against the intellect to try to mask the problem of naturalizing the epistemology of mathematics with philosophical razzle-dazzle. Superficial worries about the intellectual hygiene of causal theories of knowledge are irrelevant to and misleading from this problem, for the problem is not so much about causality as about the very possibility of natural knowledge of abstract objects. (Hart (1977), 125–6)

If the causal theory is not the problem, then attacking the causal theory doesn't help. But if the causal theories of knowledge and reference are removed from the premisses of the anti-Platonist arguments, what will take their place?

The idea that the causal theories are not the problem gains support from the historical facts: the causal theory of knowledge has gradually lost favour in the years since the appearance of Benacerraf's article, while the sentiment that there is a persuasive Benacerraf-style argument against Platonism remains strong. One might try to pin down this new argument by combing the contemporary epistemological literature for a descendant of the causal theory that could take its place in the first premiss. The best candidate would be reliabilism: for my justified, true belief to count as knowledge, it must be generated by a reliable process. Depending on how this account is spelled out, it may or may not involve a causal constraint strong enough to do the causal theory's job in the argument against Platonism.[23]

I won't go into the details of current reliabilist epistemology because in fact I don't think the force of Benacerraf-style thinking depends on this particular philosophical epistemology any more than it depended on the causal theory. To the extent that these are

[21] See Benacerraf (1973), 413.

[22] Hart makes this point (1977, p. 124).

[23] In Maddy (1984*a*), I indicate that reliabilism might not do this job. Casullo (forthcoming) gives a more refined account that suggests otherwise.

intended as a priori philosophical theories of what knowledge or justification consists in, any broad sceptical conclusion based on them—e.g. that mathematics is not a science—errs against the tenets of epistemology naturalized. To the extent that reliabilism and the rest are proposals for naturalized accounts of what knowledge is, given the overwhelming evidence in favour of mathematical knowledge, they will not last long as parts of our best theory if they purport to rule it out.[24] Indeed, for all we know, from the naturalized perspective, the very notions of knowledge and/or justification might be ultimately dispensable. But for all that, I think a Benacerraf-style worry would remain.

To see this, recall the discussion of truth and reference in the third section of Chapter 1. The current debate between supporters of correspondence and disquotational truth suggests that even if the correspondence theorist's substantive notions of truth and reference turn out to be scientifically dispensable, there will remain a problem of explaining the reliability of an expert's beliefs about the field of her expertise. Now I want to add a similar assessment of the situation in the theory of knowledge. Even if reliabilism turns out not to be the correct analysis of knowledge and justification, indeed, even if knowledge and justification themselves turn out to be dispensable notions, there will remain the problem of explaining the undeniable fact of our expert's reliability. In particular, even from a completely naturalized perspective, the Platonist still owes us an explanation of how and why Solovay's beliefs about sets are reliable indicators of the truth about sets.[25]

The nominalist Hartry Field, realizing that the causal theory of knowledge is something 'almost no one believes any more',[26] ends up rephrasing the Benacerraf worry in very similar terms: '... Benacerraf's challenge ... is to provide an account of the mechanisms that explain how our beliefs about these remote entities can so well reflect the facts about them.' Field combines this with the traditional Platonist's conception of mathematical entities:

> The relevant facts about how the platonist conceives of mathematical objects include their mind-independence and language-independence; the fact that they bear no spatio-temporal relations to us; the fact that they do

[24] See Burgess (1983), 101.

[25] R. M. Solovay is one of our leading contemporary set theorists.

[26] All the quotations in this paragraph come from Field (1989), 25–7.

not undergo any physical interactions (exchanges of energy-momentum and the like) with us or anything we can observe; etc.

From these two premisses, he draws a guarded pro-nominalist conclusion:

> The idea is that *if it appears in principle impossible to explain this*, then that tends to *undermine* the belief in mathematical entities, *despite* whatever reason we might have for believing in them. . . . Like Benacerraf, I refrain from making any sweeping assertion about the impossibility of the required explanation. However, I am not at all optimistic about the prospects of providing it.

Here we have a statement of the problem that makes no appeal to theories of truth, reference, justification, or knowledge. It simply, naturalistically, asks for an explanation of a purported fact.

Of course, there is more to it than this. If this new version of the Benacerrafian syllogism is to share the anti-Platonistic moral of the original, there must be strong reasons to suppose that the required explanation will not be forthcoming. Field's formulation surely implies this, and in my discussion of truth, I also suggested that whatever difficulties the theory of reference was supposed to cause the Platonist, the obstacles to explaining reliability would be similar, perhaps identical. But on the surface, this new argument hardly seems to share the knock-down conclusiveness of its predecessor. Given that the first premiss, the replacement for the causal requirement, includes no clause that stands in explicit contradiction to the traditional Platonistic assumptions of the second, we must ask why the task of providing the required explanation should still seem so daunting.

Obviously, what we are up against here is another, less specific, version of the same vague conviction that makes the causal theory of knowledge so persuasive: in order to be dependable, the process by which I come to believe claims about *x*s must ultimately be responsive in some appropriate way to actual *x*s. And, invoking the second premiss, nothing can be responsive to non-spatio-temporal, unchanging, acausal, unobservable Platonic entities. How, then, can Solovay's reliability be anything more than a fluke? How can it possibly be explained?

I won't try to make this vague conviction any more definite, nor will I try to refute it. But I do want to point out it is not, by itself, enough to cause a problem for traditional Platonism. Even if our most basic reliable beliefs, for example perceptual beliefs, are directly conditioned by the objects of those beliefs, many other, less basic beliefs are inferred from these. A physicist, for example, needn't see, or causally interact with, a molecule of water on the other side of the moon in order to reliably believe that it has a certain structure; she need only have good reason to believe that there is water on the other side of the moon and a well-established theory of the structure of water molecules. In other words, any reasonable theory of reliability will have to allow for various forms of inference as reliable belief-forming mechanisms.

Why then couldn't Solovay's beliefs about sets also be reliably inferred? Of course, many of them are, assuming deduction is reliable, but what about the axioms from which these are deduced? In fact, this possibility is rarely considered; underlying the persuasiveness of all these Benacerraf-style arguments—from the original one based on the causal theory of knowledge to the current one based on the need for an explanation of reliability—is the unspoken assumption that some mathematical beliefs are not inferred. This omission is especially glaring given that one prominent form of Platonism, namely Quine/Putnamism, does treat all mathematics as inferred: mathematics is a collection of highly theoretical hypotheses, justified by their indispensable role in science. As this sort of hypothetical inference might well qualify as a reliable process, Quine/Putnamism should at least be considered as a possible reply to Benacerraf-style worries.[27]

But as I say, it often isn't,[28] a fact for which I offer this explanation. Mathematicians aren't the only ones swayed by the pre-theoretic realism described at the beginning of Chapter 1. Many of us tend to think of mathematics, not as a highly theoretical adjunct to physical science, but as a science in its own right, with its own subject matter and its own methods. On this view, mathematics is parallel to, not subservient to, natural science, so it is natural to suppose that Platonistic epistemology should run parallel to scientific epistemology, and from this it follows that some mathematical beliefs should be basic and non-inferential, just as

[27] Of course, as indicated in ch. 1, sect. 4, above, I don't think Quine/Putnamism offers an acceptable account of mathematical knowledge (or the reliability of mathematical beliefs), but my point here is that Benacerraf and those who cite him rarely even consider it.

[28] Field (1989), 28–30, is an exception. He rejects a Quine/Putnam solution for some of the reasons rehearsed in ch. 1, sect. 4, above.

some scientific beliefs are. Furthermore, as the most fundamental belief-forming mechanism in physical science is perception, the corresponding faculty in mathematical science is expected to be 'perception-like', and hence, most likely causal.[29] And here the seeming impossibility arises.

But if this pre-theoretic science/mathematics analogy lies behind the stubborn Benacerraf-style worries of some philosophers, I don't want to suggest that it accounts for all. Even an advocate of a Quine/Putnam Platonism, who holds that all mathematical beliefs are gained by the reliable process of inference to the best explanation, might worry all the same because she holds that all explanations are ultimately causal. For another, the role of the science/mathematics analogy might be played instead by a strong form of physicalism that requires every legitimate entity to be part, as Armstrong puts it, of 'a single, all-embracing spatio-temporal system'.[30] And there are doubtless other possibilities. Thus, I think it's best to see these serious philosophical worries about Platonism as a syndrome with more than one aetiology.

One last question: does the nagging Benacerraf-style problem for Platonism constitute an argument in favour of nominalism? Of course, many anti-Platonists have claimed that it does, but John Burgess strenuously disputes this.[31] From a naturalized perspective, epistemology is a descriptive and explanatory enterprise; its goal is to describe and explain the belief-forming mechanisms of human knowers. So, Burgess considers the actual practices of scientists. Here we find well-confirmed affirmations of mathematical knowledge and no attributions of causal powers to mathematical entities. Thus, a causal requirement on knowledge of the sort enshrined in Benacerraf's first premiss simply doesn't turn up in the descriptive phase of epistemology naturalized. He concludes that 'a causal

[29] I'm assuming here that perceptual beliefs aren't inferential, in particular, that they aren't inferred from sense data. This doesn't rule out 'inference to the best explanation' accounts of perception (e.g. Harman (1973), ch. 11, or Gregory (1970; 1972)) because there the inferences are from states of the nervous system to beliefs, and I'm reserving the word 'inference' for inferences from beliefs to beliefs.

Also, there are those who hold that perception needn't require causation (Kim (1977)), but I see no need to quibble about this. The issue is whether or not mathematical objects can participate in interactions suitably similar to the participation of my hand in the formation of my perceptual belief that there is a hand before me when I look at it in good light. See Grice (1961), and sect. 2 below.

[30] Armstrong (1977), 149. I'll return to this idea in ch. 5, sect. 1, below.

[31] See Burgess (forthcoming *b*).

criterion for knowledge [is] problematic, whether regarded as part of a proposed analysis of the meaning of "know" or as part of a proposed analysis of scientific standards of justification' (Burgess (forthcoming *b*)). Presumably the same could be said for the requirements on reliability implicit in Field's version of the argument, and even for the various other assumptions on which I have suggested that Benacerraf-style worries might rest.

Field naturally sees the situation somewhat differently. He doesn't claim that Benacerraf-style worries provide a conclusive argument against Platonism; he recognizes that they do nothing to disarm the otherwise powerful arguments for Platonism:

> Of course, the reasons for believing in mathematical entities (in particular, the indispensability arguments) still need to be addressed, but the role of the Benacerrafian challenge (as I see it) is to raise the cost of thinking that the postulation of mathematical entities is a proper solution, and to thereby increase the motivation for showing that mathematics is not really indispensable after all. (Field (1989), 26)

That is Field's plan: to undermine the indispensability arguments by showing how mathematics could be useful in applications without being true. This project requires that science be rewritten in nominalistically acceptable terms, *contra* Putnam's claims, and that the use of mathematics in nominalized science be justified nominalistically, without appeal even to Hilbert-style finitistic metamathematics.

If this were possible,[32] Field would be presenting an alternative overall theory, according to which there is no mathematical knowledge, but the use of mathematics in science is nevertheless justified. The advantage of this picture over the Platonist's is that it doesn't leave a puzzling open question in the psychological part of our theory about how people come to have reliable beliefs about Platonic entities. Thus I take it Field would be arguing, in good naturalized form, that his overall theory of the world is better than the Platonist's. At this point, then, the Benacerraf-style problem would become an argument in favour of nominalism.

Even then, though, I suspect Burgess would doubt that Field's theory is actually preferable. After all, it requires a revision in the

[32] Numerous difficulties with Field's project have emerged. For a sampling, see Malament (1982), Resnik (1985*a*, *b*), Shapiro (1983*b*), Burgess (1984), and Detlefsen (1986), ch. 1. For my own discussion, see ch. 5, sect. 2, below.

standard canons of scientific practice to balance a worry generated by a vague and surely less well-established psychological conviction about the nature of reliable processes. Perhaps he would argue that no gain in the psychological portion of our theory could justify the rejection of otherwise effective scientific methods. At this point, the form of Field's reply would depend on the actual details of his proposed nominalist theory.

This is an important debate, one I won't attempt to resolve, but I do want to point out that, despite appearances, it is irrelevant to my project. If Burgess is right, his arguments would undermine the effectiveness of Benacerraf-style worries as justifications for nominalism. However welcome this conclusion might be, it does nothing to take the compromise Platonist off the epistemological hook. By rejecting pure Quine/Putnamism, by embracing some version of Gödel's science/mathematics parallelism, the compromise Platonist incurs the very real debt detailed in the last chapter: within the bounds of epistemology naturalized, she owes a descriptive and explanatory account of mathematical knowledge (or mathematical reliability) that does justice to the actual practice of mathematics, an account of both intuition and other peculiarly mathematical justifications. And to provide such an account is to meet the Benacerraf-style worries head on, whether or not they constitute an effective argument for nominalism.

So, despite all this talk about the exact (or inexact!) nature of the causal premiss to the Benacerrafian argument, it is not my plan to attack the problem at that point. Instead, I intend to reject the traditional Platonist's characterization of mathematical objects; I will bring them into the world we know and into contact with our familiar cognitive apparatus. An account of our 'perception-like' connection will be the goal of this chapter. To give a taste of how it will go, let me return to the (convenient, but inessential) language of the causal theories.

Consider again the initial baptism of gold: the baptist stands in front of an array of samples, looks at them, and declares that these and things like them are gold. An analogous baptism of sets would go like this: our baptist, at her desk, declares, 'These three things—the paper weight, the globe, and the inkwell—taken together, regardless of order, form a set', or 'The individual books on this shelf, taken together, in no particular order, form a set.' In this way, she isolates samples of the kind 'set', and the word then refers to the kind of which these samples are members.

The obvious objection is that, while the gold-dubber causally interacts with her samples, the set-dubber causally interacts only with the members of her samples. Here the Platonist might respond:[33] the extent of the causal interactions of both the gold-dubber and the set-dubber is something like light bouncing off certain objects and bringing about some retinal changes. In the case of the gold-dubber, strictly speaking, the interaction is actually with the front surface of a time slice of the sample. In other words, the thing whose kind we count the baptist as having dubbed is not the thing with which the dubber has causally interacted; her interaction is only with a fleeting aspect of the temporally extended sample. Similarly, the set-dubber has only aspects of her sample sets within her causal grasp. But, if the interaction of the gold-dubber with an aspect of her samples is enough to allow her to pick out a sample and dub a kind, why shouldn't the set-dubber's interaction with an aspect of her samples accomplish those same things? The Platonist could argue that the relation of element to set is no more objectionable than the relation of fleeting aspect to temporally extended object.[34]

If an argument of this sort could be filled in, then it seems the Platonist might adopt a causal theory of reference after all. And since the bare causal interactions described form the basis for a perceptual connection between the baptist and her samples, a causal theory of knowledge is within reach as well. Both causal theories require more than the mere causal interaction—the bouncing of light rays off objects and onto retinas—they require that the baptist and the knower *perceive* a physical object. That these two are not the same is clear from experiments on patients, blind from birth, whose sense organs are restored to perfect operating condition, but who cannot properly perceive the objects around them.[35] The Platonist's hope is that an account of what

[33] That an argument of this sort might be available to the Platonist was first suggested to me by John Burgess.

[34] Of course, the set-dubber's interaction is also with mere fleeting aspects of the members of her sets, so the relation between what she interacts with, and what's kind she dubs, is the composition of the aspect/object and the element/set relations. This added complexity could be eliminated by imagining that the set-dubber uses sets of aspects, rather than sets of objects, as samples, but a more reasonable course would be to assume that if both the relations in question are legitimate, then so is their composition.

[35] See Hebb (1949), ch. 2.

makes one pattern of sensory stimulation into a perception of a physical object might also provide an account of what makes another pattern of stimulation into a perception of a set of physical objects.

2. Perception

The question is what bridges the gap between what is causally interacted with and what is perceived, and the hope is that something like what does the bridging in the case of physical object perception can be seen to do the same job in the case of set perception.[36] Notice that this way of putting the problem already assumes that we do in fact perceive physical objects, as opposed to sense data, or percepts, or representations of some kind. In the theory of perception, this is called 'direct realism'. Psychologists of perception are generally direct realists in this sense,[37] and though some philosophers are still inclined to debate the issue, I'll rely implicitly on the persuasive counter-arguments to be found in the literature.[38]

What we want here is a strong sense of 'perceives' that rules out illusions; what is perceived in this sense is really there. But that isn't all. We also insist that, for example, a hiker doesn't perceive a leaf-dwelling insect on a bush she passes if the bug blends too perfectly with its surroundings for her to distinguish it from them. In such a case, even though light from the bug registers a pattern on the hiker's retina, she gains no beliefs about it. It isn't enough, however, to require that she gain beliefs in order to perceive, or even that she gain beliefs visually, because it is possible to gain beliefs, even visually, that clearly don't count as perceptual beliefs about the bug; she could, for example, come to believe there is a Ceylonese leaf insect on the bush by reading a sign. This belief isn't a perceptual belief that there is a Ceylonese leaf insect on the bush because it doesn't involve it looking to the hiker (in a phenomenal or non-metaphorical sense) as if there is a bug on the bush. For

[36] I will concentrate here on visual perception. Naturally the blind can know and refer as well as the sighted, but I make the customary philosophical assumption that what's true of vision can be adapted to the other senses.

[37] For history, see Hebb (1980), 19. For agreement from various psychological camps, see J. Gibson (1950), 26–7, Neisser (1976), 16, and Gregory (1972), 220.

[38] See Pitcher (1971), ch. 1, or the references cited in Machamer (1970), § II.

perception, then, we require that the perceiver gain appropriate perceptual beliefs.

But this still isn't enough. Consider an example analogous to Gettier's: an illusionist arranges a system of mirrors so that it looks to Steve as if there is a tree in front of him. Suppose that the actual source of the image Steve sees is a tree behind him. Suppose, finally, that there is in fact a tree in front of him, where he sees the illusion, though that tree is hidden. Here Steve gains a true perceptual belief—there is a tree in front of me—but it doesn't count as a perception that there is a tree in front of him because its causal genesis is faulty. Clearly, we must insist that whatever makes the belief true—in this case, the tree in front of Steve—be responsible for his belief in some appropriate way. Paul Grice's solution is to insist that the thing perceived must play the same sort of role in the causation of the perceiver's perceptual state as my hand plays in the generation of my belief that there is a hand before me when I look at it in good light.[39] In sum, then, for Steve to perceive a tree before him is for there to be a tree before him, for him to gain perceptual beliefs, in particular that there is a tree before him, and for the tree before him to play an appropriate causal role in the generation of these perceptual beliefs.[40]

Notice that the content of a perceptual belief state is extremely rich and varied. For Steve to acquire the perceptual belief that there is a tree before him, he must also acquire a great variety of other perceptual beliefs, depending on the occasion, such as, that the tree is roughly so big, so far away, that it is in leaf, swaying in the breeze, and so on.[41] Such beliefs are non-inferential,[42] and not necessarily conscious or linguistic.[43] When the various components of a perceptual belief state arise as a body, on a given occasion, they often influence each other non-inferentially, as, for example, a belief about the identity of an object can influence perceptual beliefs about its shape and size, and obviously, vice versa.

[39] See Grice (1961).

[40] This account is a crude version of Pitcher's (1971), ch. 2. It is similar to Armstrong's (1961), and to various psychological theories of perception as information acquisition, like Gregory's (1972). Pitcher and Armstrong insist that belief content exhausts the perceptual state, while Goldman and others aren't so sure. (See Goldman (1977), § 6.) For our purposes, this issue is beside the point.

[41] See Pitcher (1971), 87–9.

[42] Except for the weak sense, noted earlier, in which they can be considered as 'inferred' from states of the nervous system.

[43] See Armstrong (1973), chs. 2 and 3.

Beliefs in general are psychological states. I assume that a person's behaviour gives good evidence for our hypotheses about her psychological state, though I won't go so far as to assert, as some philosophers would, that being in a certain psychological state simply is behaving (or being disposed to behave) in certain ways. If this behaviouristic position is incorrect, then behavioural evidence is not conclusive, but it is still often the best available. Sometimes, there might also be introspective evidence for or against the claim that a person is in a given psychological state, and if there is a correspondence (not necessarily the identity) between psychological states and brain states, as I suppose there is, then neurophysiological evidence would also be relevant.[44]

The question before us now is this: how do we manage to perceive physical objects? Assuming[45] that to have a concept is to have the capacity for beliefs of a certain sort, we can rephrase the question: how do we come to have the concept of a physical object? Psychologists and neuropsychologists have produced various results and theories to answer the question of how conceptual elements enter human perceptual states. I'll review some of their work here before returning to the philosophical issues involved.

There is considerable experimental evidence that the ability to perceive a primitive distinction between a figure and its background is inborn in humans and many laboratory animals.[46] The structure of the retina is probably responsible for the presence of this conceptual information in the human perceptual state, much as it is in the frog. Warren McCulloch and his co-workers have isolated various structures in the frog's retina which send impulses to the frog's brain only under certain complex sets of conditions, independent of the level of general illumination, for example, in the presence of sharp boundaries between relatively light and relatively dark patches, or dark areas with curved edges, or movement of such edges. In fact, one fibre

> responds best when a dark object, smaller than a receptive field, enters that field, stops, and moves about intermittently thereafter. The response is not affected if the lighting changes or if the background (say a picture of grass

[44] Of course it will take some substantial progress in neuroscience before this is a real possibility, but my point is that it isn't ruled out a priori.

[45] With Armstrong (1973), ch. 5, § 1.

[46] See Hebb (1949), 19–21.

and flowers) is moving, and is not there if only the background, moving or still, is in the field. Could one better describe a system for detecting an accessible bug? (Lettvin *et al.* (1959), 254)

As might be expected, the researchers came to think of these fibres as 'bug-detectors', and the frog's behaviour certainly suggests that this mechanism enables it to acquire perceptual beliefs about nearby bugs. Similar mechanisms in humans are probably responsible for perceptual beliefs concerning figure and background, and perhaps some concerning distance and size.[47]

But beyond this fairly simple level, the evidence indicates that the capacity to acquire perceptual beliefs of the familiar sort is not present at birth.[48] Psychologists talk of a phenomenon called 'identity' in perception. A figure is said to be seen with identity when it appears similar to some other figures but not to others, when it is seen as falling into some categories and not in others, when it is easily recalled, recognized, or named. When I see a triangular figure, for example, I automatically see it as more like other triangles than like squares, I can recall it, recognize it, and call it and other similar figures 'triangles'. In the terminology we've adopted here, I acquire the perceptual belief that there is a triangle before me. Experiments on newly sighted human patients who had been blind from birth, and on chimpanzees raised in total darkness, demonstrate that the capacity to acquire such a belief—what we've also called having the concept of a triangle—is present only after considerable perceptual experience. For example,

Investigators (of vision following operation for congenital cataract) are unanimous in reporting that the perception of a square, circle, or triangle, or of sphere or cube, is very poor. To see one of these as a whole object, with distinctive characteristics immediately evident, is not possible for a long period. The most intelligent and best-motivated patient has to seek corners painstakingly even to distinguish a triangle from a circle . . . A patient was trained to discriminate square from triangle over a period of 13 days, and had learned so little in this time 'that he could not report their form without counting corners one after another . . . And yet it seems that the recognition process was beginning already to be automatic, so that some day the judgement "square" would be given with simple vision,

[47] See e.g. Bower (1966).

[48] This has little to do with the philosophical controversy over innateness because even their defenders admit that something sensory is needed to 'draw out' or 'awaken' innate ideas.

which would then easily lead to the belief that form was always simultaneously given'. (Hebb (1949), 28, 32)

Similar results were obtained with the chimpanzees.

Given that a capacity as simple as the ability to see a triangle as more like another triangle than like a square is the product of considerable sensory experience, it is to be expected that so complex a talent as that of seeing a series of different patterns as aspects of one thing—that is, as a sequence of views of one physical object—is not present at birth. This expectation is substantiated by the experiments of Jean Piaget and his colleagues.[49] The child's ability to acquire perceptual beliefs about physical objects, as judged from behaviour, develops between the ages of one and eighteen months. At the beginning of this period, the child's world is a welter of isolated incidents. Then

the behavior of the child begins to be centered on *objects*; but to him there is no objective reality—no general space or time, no permanence of objects. There are only *events*—i.e. components of the child's own functioning. When an object in his field of vision disappears, it ceases to exist. (Phillips (1975), 28)

Some months later, objects begin to enjoy a sort of permanence:

there is a shift in the child's conceptualization from object reality dependent on his own actions to object reality dependent on the surround. The result is a kind of 'context-bound object permanence'. (Phillips (1975), 38)

At this stage, the object is associated with a particular location; the child expects to find it there even when it is clearly hidden in a new location. Similarly, we find that

the infant does not realize that a moving object is the same object when it becomes stationary . . . [or] that a stationary object that begins to move is still the same object after it starts moving. (Bower (1982), 206)[50]

The ability to distinguish the object by features such as size, shape, and colour, in addition to its location or trajectory, is a major

[49] See Piaget (1937) and Phillips (1975).

[50] Bower and his colleagues would dispute some details in the Piagetian account of the development of the object concept, but the general idea of a non-trivial developmental period is all that matters here. Even psychologists who play down the role of learning in perception agree that a development of this general sort takes place. See E. Gibson (1969), ch. 16.

development, followed by the ability to distinguish objects that share a common boundary and a stepwise improvement in searching behaviours. By the end of the developmental period, the child possesses our familiar concept of an independently existing physical object and is fully capable of acquiring perceptual beliefs about them.

Supposing then, as the evidence suggests we should, that the concept of a triangle or of a physical object—that is, the ability to acquire beliefs, including perceptual ones, about such things—is acquired over some period of time, what can be said about how this is accomplished? Presumably some change must take place within the brain, some alteration in the neural structuring, that enables us to see things with identity, that fills the gap between the pattern of light on our retinas and the perceptual beliefs we acquire. To fix our ideas, I'll sketch a particular neurophysiological theory of what goes on in this development, a theory due to Donald Hebb.[51] I doubt that the philosophical morals I'll eventually draw actually depend on the correctness of exactly this scientific theory, but in an area of abstract epistemology, I find it reassuring, even useful, to have at least one fairly specific example of how a naturalistic story of our knowledge might go.

That part of the brain involved in perception, the visual cortex, can be divided into four layers, and the pattern of retinal stimulation is topologically equivalent to the pattern of activity in the first of these only. After that, topological characteristics are not preserved, and the neural connections between the inner layers, and between these and the outer layer, are so complex as to seem random. Excitation of a small part of the initial layer can stimulate widely separated areas in all three inner layers; widely separated parts of the initial layer can stimulate neighbouring cells in the inner layers; inner areas can stimulate other inner areas and even outer ones. For any one cell to fire and pass along its excitement to the cells connected with it, it must be stimulated by many other cells simultaneously. But any cell in the visual cortex is connected with many, many others, so the firing of a number of cells in the first layer will usually result in a convergence of sufficient stimulation for firing on a number of cells in various other layers, if for statistical reasons alone. Thus any visual stimulus creates a

[51] See Hebb (1949), esp. chs. 4 and 5, and (1980), ch. 6.

veritable hum of activity throughout the visual cortex, and the hums corresponding to different retinal stimuli are globally the same. The question, then, is how all this activity is organized.

To answer it, Hebb makes a simple theoretical assumption, namely, that if one cell repeatedly plays a role in the firing of another, then a change takes place increasing the first cell's efficacy in firing the second. He suggests that this phenomenon could be produced by the growth of synaptic knobs, but for theoretical purposes, any plausible mechanism will do. Now suppose that someone unfamiliar with triangles fixes her gaze on the apex of a given triangular figure. This generates a certain pattern of stimulations which recurs as long as she stares at that corner, and occurs again every time she looks at it. As a result of this repeated exposure, groups of cells at various cortical levels are repeatedly efficacious in firing one another (at convergences of various kinds). As a result, it becomes easier for these interconnected cells to fire one another, they become more interdependent, eventually forming what Hebb calls a 'cell-assembly'. It will respond to the apex of any similar triangle.

Analogous processes naturally result in assemblies for the base angles of the triangle, but this leaves open the question of how the triangle as a whole, as a unit, is perceived with identity. Behavioural experiments suggest that acquiring the ability to recognize triangles depends essentially on successive eye movements and fixations on various parts of the figure. The eye is especially inclined to trace lines because all the movement-inducing peripheral stimulations are urging it in one direction. Thus angles are frequent fixation points, lying as they do at the intersection of straight lines, and when the eye is fixated on one angle, it is stimulated to move towards another.

Because of the numerous interconnections between neurons in various parts of the visual cortex, cells in one assembly happen by chance to be connected with various cells in the other two. When the motor stimulations described above induce repeated movement from one angle to another, the neurons in these chance connections become increasingly efficacious, and the cell-assemblies for the three corners are integrated into a second-order cell-assembly. The individual angle assemblies can still work independently, so actual perception of a triangle involves what's called a 'phase sequence' of excitations of the corner assemblies and the integrated assembly. If the corner assemblies are *a*, *b*, and *c*, and the integrated assembly *t*, then a phase sequence is something like *a–b–t–a–c–t–c–t–b*.[52]

Once an integrated cell-assembly of this sort has been formed, looking at a triangle will cause it to reverberate for half a second or more. This represents a considerable gain both in organization and in duration over the random hum of activity brought about by the same visual stimulation before the formation of the assembly. This longer, repeatable trace should persist long enough to allow the structural changes required for long-term memory. In other words, the cell-assembly is what permits the subject to see a triangle with identity, to acquire perceptual beliefs about it: it is a triangle-detector in much the same sense as the fibre located by McCulloch is a bug-detector; it provides the subject with her concept of triangle.[53]

The ability to perceive physical objects is not unlike the ability to perceive triangular figures, though it is more complex. The trick is to see a series of patterns as constituting views of a single thing. Just as the ability to see triangles develops over time, through a painstaking process of seeking out corners and comparing one triangle with another, the ability to see continuing physical objects develops over a period of experience with watching and manipulating them. The close resemblance in the structure of the learning processes suggests that what is involved in the physical object case is just a more elaborate version of the cell-assembly, in particular, the development of higher-order cell-assemblies which respond to a

[52] This account is based on Hebb (1949), chs. 4 and 5. Hebb (1980) cites evidence suggesting that the basic assemblies may respond to the sides of the triangle rather than its angles (see pp. 89, 98–102). This modification doesn't affect the upshot of my discussion. The 1980 work also summarizes new supports for Hebb's theory discovered in the years since 1949 (see pp. 98–100), and develops various applications of the view, including a fascinating account of scientific problem-solving (pp. 119–21).

[53] Philosophers of mind sometimes worry about a blanket objection Dennett has raised to attempts to locate concepts in the neurons: 'Suppose your "grandmother" neuron died; not only could you not *say* "grandmother", you couldn't *see* her if she was standing right in front of you . . . you would have a complete cognitive blind spot . . . Nothing remotely like that pathology is observed, of course, and neurons malfunction or die with depressing regularity, so . . . theories that require grandmother neurons are in trouble' (1978, p. xiii). I don't think Hebb's theory requires anything so specific as a grandmother neuron, but it does posit shape and other general detectors. Hebb (1949) already contains a response to objections like Dennett's: 'The assembly is thought of as a system inherently involving some equipotentiality, in the presence of alternate pathways each having the same function, so that brain damage might remove some pathways without preventing the system from functioning . . .' (p. 74).

series of aspects of a single object, and fire, as before, in a complicated phase sequence.[54] When an object stimulates a phase sequence of such assemblies, it participates in the generation of the subject's perceptual belief state in the appropriate causal way, in the way my hand participates in the generation of my belief that there is a hand before me when I look at it in good light.

Crudely put, human beings develop neural object-detectors which allow them to perceive independent, continuing physical objects. It is these complex cell-assemblies that bridge the gap between what is interacted with and what is perceived. The object I perceive on a given occasion, or more precisely, the front side of a time slice of that object, is causally responsible only for the pattern of retinal stimulations, while the unifying concept of a familiar physical object is contributed by my physical object-detector. The presence of the object-detector, in turn, is partly the result of the structure of my brain at birth (conditioned by the evolutionary pressures of the environment on my ancestors) and partly the result of my early childhood experiences with physical objects. All this, while undeniably complex, is still naturalistic, and causal.

Let me now return to the subject that inspired this detour into the theory of perception in the first place, namely my claim that we can and do perceive sets, and that our ability to do so develops in much the same way as our ability to see physical objects. Consider the following case: Steve needs two eggs for a certain recipe. The egg carton he takes from the refrigerator feels ominously light. He opens the carton and sees, to his relief, three eggs there. My claim is that Steve has perceived a set of three eggs. By the account of perception just canvassed, this requires that there be a set of three eggs in the carton, that Steve acquire perceptual beliefs about it, and that the set of eggs participate in the generation of these perceptual beliefs in the same way that my hand participates in the generation of my belief that there is a hand before me when I look at it in good light.

This claim will doubtless elicit a clamour of objections, and even if it doesn't, it needs elucidation, so let me begin at the beginning, with the assertion that there is a set of eggs in the carton in front of Steve. The simplest way to resist this idea is to deny that there are sets. To this I reply with a version of the Quine/Putnam arguments

[54] See Hebb (1980), 107.

sketched towards the end of Chapter 1—mathematical entities are indispensable for our best theory of the world—supplemented by the observation that our best theory of mathematical ontology is that (at least some)[55] mathematical entities are sets.

A second mode of resistance is to join the traditional Platonist in denying that sets have location in space or time. But notice: there is no real obstacle to the position that the set of eggs comes into and goes out of existence when they do, and that, spatially as well as temporally, it is located exactly where they are. A set of higher order, like the set consisting of the set of eggs and the set of Steve's two hands, would again be located where its members are, that is, where the set of eggs and the set of hands are, which is to say, where the eggs and hands are. In this way, even an extremely complicated set would have spatio-temporal location, as long as it has physical things in its transitive closure.[56] And any number of different sets would be located in the same place; for example, the set of the set of three eggs and the set of two hands is located in the same place as the set of the set of two eggs and the set of the other egg and the two hands.[57] None of this is any more surprising than that fifty-two cards can be located in the same place as a deck. In any case, I hereby adopt this view as part of set theoretic realism. On some terminological conventions, this means that sets no longer count as 'abstract'. So be it; I attach no importance to the term.[58]

More controversial is the second part of the claim that Steve sees a set: the contention that he gains perceptual beliefs about the set of eggs, in particular, that this set is three-membered. Let me break my defence of this idea into two parts. First, and least controversially, I contend that the numerical belief—there are three eggs in the carton—is perceptual,[59] that is, that it looks to Steve, in a non-metaphorical sense, as if there are three eggs there. There is

[55] As indicated in ch. 1, sect. 1, I'm not assuming that all mathematical entities are sets, but our standard mathematical theories indispensably refer to sets of points, sets of numbers, etc. See ch. 3 below for the relationship of the familiar natural and real numbers to sets.

[56] The transitive closure of a set consists of its members, the members of its members, the members of the members of its members, and so on. For a formal definition, see Enderton (1977), 178.

[57] Set theorists will notice that this means there are proper class-many sets located where any physical thing is. (They stack neatly!) Pure sets have no location. For more on pure sets, see ch. 5, sect. 1, below.

[58] See Katz (1981), 207 n. 29. Parsons uses the term 'quasi-concrete' for mathematical objects of this sort. See his (1983*a*), 26.

[59] See Kim (1981) for a similar claim.

empirical evidence, based on reaction times, that such beliefs about small numbers are non-inferential.[60] Furthermore, this belief about the number of eggs can non-inferentially influence and be influenced by other clearly perceptual beliefs acquired on this occasion; for example, the welcome fact that there are enough eggs for the recipe can make the eggs themselves look larger.[61] This particular perceptual belief about the number of eggs is thus part of a rich collection of perceptual beliefs acquired on this occasion, beliefs about the size and colour of the eggs, the fact that two eggs can be selected from among the three in various ways, the locations of the eggs in the nearly empty carton, and so on.

So far so good. Now let me take up the question of whether or not this perceptual belief is a belief about a set. What is a numerical belief about, after all? The easiest answer would be that Steve's belief is about the eggs, the physical stuff there in the carton, but Frege long ago demonstrated the inadequacy of that response.[62] The trouble is that the physical stuff in the carton has no determinate number property: it is three eggs, but many more molecules, even more atoms, and only a quarter of a carton of eggs. For a given mass of physical stuff, there is no predetermined way that it must be divided up, and without this, there is no determinate number property. So the physical stuff by itself cannot be three.

If not the physical stuff making up the eggs, then what is the subject of a number property? Some would say, 'the eggs', meaning by this the physical stuff as divided up by the property of being an egg, what I'll call an 'aggregate'.[63] Frege's answer is that a numerical statement is a statement about a concept, for example, the concept 'egg in the carton in front of Steve'. Others might choose the extension of Frege's concept, that is, what is usually

[60] See Kaufman *et al.* (1949).

[61] For another example of such non-inferential influence, suppose a majority vote from a three-membered panel will defeat your motion. When two panellists raise their hands to vote no, your numerical belief that there are two of them (enough to dash your hopes) may well influence your perception of their facial expressions (how malevolent they look!). If only one had voted negatively, she might only have looked dense.

[62] Frege (1884), §§ 22–3.

[63] Some would say that 'there are three eggs in the carton' is properly analysed as saying 'there is an x, there is a y, and there is a z, all distinct, such that x is an egg in the carton, y is an egg in the carton, z is an egg in the carton, and anything that's an egg in the carton is either x or y or z'. I take this to be a variant of the aggregate view: some physical stuff is divided up in a certain way.

called the 'class' of eggs in the carton.[64] And the set theoretic realist opts for the set of eggs in the carton.

But on what grounds? If Steve is supposed to see a set of eggs, shouldn't the set theoretic realist hold that he can see, for example, that it is a set and not an aggregate? Attractive as this move might seem, I think it is not correct. Notice that the various candidates for the bearer of the number property—the set, the aggregate, the concept, the class—have their most basic properties, any properties that might count as perceptual, in common; for example, they all have subcollections (e.g. the egg-stuff under a more exclusive property), they are all capable of combination (e.g. the disjunction of two concepts), and so on. These similarities are what makes them all potential candidates for number-bearing. The properties that separate them are theoretical properties, like extensionality. Asking Steve to look and see whether he's perceiving a set or an aggregate is like asking him to look and see whether the egg is solid or mainly an empty space littered with atoms.

What I'm getting at is this: the amount we know about things by perception is very limited. About physical objects, for example, we know little more than that they are, in Hebb's words, 'space-occupying and sense-stimulating *something*[s]'.[65] Beyond that, the bulk of our knowledge about them is theoretical: that they are made up of atoms, of this and that sort, arranged in such-and-such a way, and so on. The same goes for sets. What we perceive is simply something with a number property, something that can be combined with others of its ilk, and so on. Nailing down this number-bearer's more esoteric properties is a theoretical matter.

So, to decide the case between sets, aggregates, concepts, classes, and whatever else, we need to look, not to our perceptual experiences, but to our overall theory of the world, and we must ask which of these is best suited to playing the role of the most fundamental mathematical entity. (Compare: deciding whether Berkeleian bundles of God's experiences or the physicist's bundles of atoms are best suited to playing the role of the most fundamental

[64] A class differs from a set in that it is essentially dependent on a property, like 'being an egg in the carton in front of Steve'. Sets, by contrast, are generated iteratively, by taking at each stage every subset of what's been generated before, regardless of whether or not the members of that subset can be singled out by some property. See ch. 3, sect 3, below.

[65] Hebb (1980), 109. I'll consider this passage in more detail in the next section.

physical entity.) On this score, sets win going away; they are extremely simple and manageable entities that form the basis for a surprisingly effective and efficient mathematical theory. In contrast, properties, on which both aggregates and classes depend, are hard to handle—no comparably flexible and complete theory is known[66] —and prone to paradox—for example, consider the property a property has when it doesn't have itself.[67] And classes are little better.[68] The elementariness of the notion of set, its ease of manipulation, and the immense success of set theory, both as a foundation for other branches of mathematics and as a mathematical theory in its own right, all help to make the set of eggs the most attractive candidate for the role of number-bearer.

I take all this to support the set theoretic realist's claim that the bearers of number properties are sets, and thus, that Steve's perceptual belief is a belief about a set. But before leaving this point, I want to call attention to the contingency of this conclusion. In its support, I depend on the idea that mathematical entities are indispensable to physical science; if they weren't, there would be no reason to include sets in our overall theory. Thus my preference for sets is contingent on the way the world is, to the extent that our best theory seems to demand them.

To appreciate the force of this fact, consider, for example, one of the fundamental differences between sets and aggregates, namely, that there are sets of higher rank—sets of sets, sets of sets of sets, and so on—while there are no aggregates of aggregates. If the physical world were simpler, allowing for a simpler physical theory with no continuous phenomena, then our overall theory might have no need for real numbers, and consequently, for sets of higher rank.[69] Since we're engaged in science fiction, we might imagine that our perceptual experiences of discrete objects in this simpler

[66] Bealer (1982), ch. 5, suggests a property theory that essentially mimics set theory, but as Anderson (1987, p. 151) points out, the plausibility of some set theoretic assumptions does not carry over to their property theoretic translations.

[67] The related 'paradoxes of set theory' present no problem for sets on the iterative conception. For example, the Russell set, the set of all non-self-membered sets, cannot be formed at any stage, because new non-self-membered sets (all sets are non-self-membered on the iterative picture) will be available at the next stage. See Gödel (1947/64), 474–5.

[68] See ch. 3, sect. 3, below.

[69] Reals, as standardly constructed, occur a few stages after ω. See Enderton's construction (1977, ch. 5). The relationship of continuous phenomena with sets of higher rank will be considered in more detail in ch. 4 and ch. 5, sect. 2, below.

world are exactly like our experiences of discrete objects in this world. Still, in the simpler world, we might have no justification for including higher ranks in our overall theory, and thus, much less justification for taking our numerical perceptions to be perceptions of sets rather than aggregates. So my claim that sets are the best candidates for the bearers of number properties depends on the fact that they are the best mathematical entities for the mathematical theory this particular world—with its continuous phenomena—requires.

Let's grant, then, that there is a set of eggs in the carton, and that Steve gains the perceptual belief that this set is three-membered.[70] For this to count as Steve's perceiving a set, only one further condition must be satisfied: the set of eggs must participate, in an appropriate causal way, in the generation of Steve's belief. Appropriate participation is exemplified by the role of my hand in the generation of my perceptual belief that there is a hand before me, and that in turn comes down to my hand's stimulation of a phase sequence of cell-assemblies. So our question is: could a set of eggs do the same?

The behavioural evidence of Piaget and his colleagues suggests that the ability to gain perceptual beliefs about sets develops in a series of stages parallel to those for perceptual beliefs about physical objects, though at a somewhat later age.[71] At the beginning of this period, a child may be able to classify objects into groups in a consistent way—say, triangles with triangles and squares with squares—but she does not correctly grasp the inclusion relation—in a group of two black squares and five black circles, the child thinks there are more round things than black. For the younger child, a set ceases to exist when its subsets are attended to.

[70] Steve needn't express his belief in this way to himself; implicit in the word 'set' is a more refined theory than most people are aware of. When I say he gains a perceptual belief about a set, I mean he gains a perceptual belief about a something with a number property, which we theorists know to be a set. Analogously, when he perceived the tree in front of him, he gained a perceptual belief with less theoretical content than a botanist could provide. Nevertheless, what he perceived was the botanist's tree.

[71] Piaget and Szemińska (1941) would put that age between seven and fourteen years. See also Phillips (1975), ch. 4. More recent work suggests that these developmental periods occur somewhat earlier, between two and a half and five years. See Gelman (1977). Once again, the details are less important than the parallel between development of the physical object and set concepts.

A similar confusion is observed in connection with the set's number properties. The younger child imagines that the number of elements in a set changes when it is rearranged, particularly when its elements are moved closer together or further apart. For older children, on the other hand, once a one-to-one correspondence between two sets has been established, the belief in their equinumerosity cannot be shaken; indeed the very question seems silly to them. Once a perceptual belief about a set is gained, the thought that the set could change its number property when its elements are moved about (barring mishap) appears preposterous.

So, just as the concept of an independent and continuing physical object is acquired in stages, the concept of a set with inclusions and a constant number property is itself gained over time, and depends on experience with groups of objects. It should be noted that the child's development of the set concept is not a linguistic achievement. Of course, children are rarely taught the word 'set', but they are taught number words, and it might be thought that their early errors are primarily verbal, and that it is verbal instruction that corrects them. The evidence is heavily against this assumption:[72]

> The major point is that the development of the concept of number begins in infancy, long before speech or formal instruction play any part. The infant is forced to generate number concepts by the requirements of its everyday activities—activities so commonplace that the fondest parent barely thinks them worthy of comment. They are worth mentioning because these are the simple beginnings from which the whole structure of mathematical thinking takes root. (Bower (1982), 250)

One must expect that the set concept, like the physical object concept, could be developed in the complete absence of language.

And how does the set concept develop? We have seen the evidence that it develops over a period of time, like the object concept, and that the determining factor in both these developments is repeated exposure to the sort of things in question. The development of the object concept is brought about by the child's experiences with various physical objects in her environment, and the set concept by experiences with sets of physical objects, for example by forming one-to-one correspondences between them, by regrouping them to form salient subsets, and so on.[73]

[72] See also Phillips (1975), 145.

[73] These manipulations are crucial. Kitcher (1983, p. 103) has suggested that my

Hebb's theory of the formation of the neural triangle-detector made essential use of the behavioural evidence that development of the ability to see triangles with identity requires repeated fixations on corners of triangular figures, eye movements from one corner to another, and even, in some cases, active seeking out of corners. Because of the behavioral similarities between this process and that leading to the development of the object concept, it is theorized that an object-detector develops in a similar way, as a result of various experiences with physical objects in the environment. Given the evidence that the set concept requires a similar developmental period involving repeated experience with sets in the environment parallel to the required experiences with triangles and physical objects, it seems reasonable to assume that these interactions with sets of physical objects bring about structural changes in the brain by some complex process resembling that suggested by Hebb,[74] and that the resulting neural 'set-detector' is what enables adults to acquire perceptual beliefs about sets.

This assumption provides a solution to another difficulty. Recall that many things—a mass of physical stuff, and many different sets—occupy the same spatio-temporal location. These things also produce the same retinal stimulation, but on one occasion the stuff is seen, on another one set, on another a different set altogether, and so on. We've all been amused by the psychologist's examples in which we see a single picture first as an undifferentiated mass, then as representing a definite number of distinct objects, or the child's puzzle in which the homogeneous jungle foliage resolves itself into a pack of ferocious beasts. Where a bookbinder sees a large set of individual books (so many perhaps that she has no perceptual access to the exact number), the encyclopaedia salesman sees three

account is inconsistent with the view of a group of perceptual theorists called the 'ecological realists'. He says that their 'general claim that the information which we gain in perception concerns transformations of the sensory array caused by events in which perceived objects participate seems to be at odds with the idea that we can acquire perceptual information about unchanging abstract objects'. I merely point out that the 'abstract' objects involved here are not unchanging; in particular, they can be moved about for purposes of more ready classification and to display one-to-one correspondences. The account in the text also seems to me in harmony with the ecological realist's emphasis on invariants (the child learns that the numerable collection is invariant under transpositions of its elements) and affordances (Steve's three-membered set of eggs affords cake-making).

[74] Hebb (1949) doesn't mention set perception, but he does consider the perception of number properties of collections to be part of his theory in Hebb (1980), 122–4.

of his rival's product; where I see a set of four shoes, you might see a set of two pairs. A microscopic image looks to me like an unorganized mess of Jackson Pollock drips, while the biologist sees three paramecia and an amoeba.

Hebb's theory provides a key to these phenomena. They involve a change in perception on the part of a single subject (the child's puzzle), a difference in perception between two roughly comparable subjects (the bookbinder and the encyclopaedia salesman), and a difference in perception between two subjects with different training (me and the biologist). The last case is easiest: perceptual development continues past childhood; the biologist acquired further perceptual abilities (further cell-assemblies) during her education in lab techniques. In the other two cases, preceding experiences and neural activity (thoughts) influence the ease with which a given pattern of stimulation will trigger a given cell-assembly: the bookbinder and the salesman have different interests; they notice different things. In the puzzle cases, some shift in our attention causes a new cell-assembly to come into play quite suddenly. If there were no such set assemblies, the phenomenon of seeing different groupings with different number properties would have to be explained by some additional organizing events occurring after the initial stimulation of the ordinary physical object assemblies. This approach is less true to the phenomenon.[75]

On this account, then, when Steve looks in the egg carton, there is a set of eggs there, he acquires perceptual belief about that set, namely, that it has three members,[76] and the set of eggs participates in the generation of his perceptual belief in the appropriate way, that is, in the way my hand participates in the generation of my belief that there is a hand before me when I look at it in good light, which is, by some aspect of the object in question causally interacting with the retina in such a way as to bring about the stimulation of the appropriate detector. In the case of sets, just as in the case of physical objects, it is the presence of a complex neural development that bridges the gap between what is causally interacted with and what is perceived.

[75] See Hebb (1980), 83–7. Also Bruner (1957), 241–4.

[76] He undoubtedly acquires other perceptual beliefs about the set of eggs at the same time, e.g. that it has various two-element subsets. Notice also that, as promised, Gettier-style cases are easily constructed; the same illusionist who set up the tree example described earlier could arrange his mirrors to create the illusion of a set where there is actually another, hidden, set.

Thus, the Hebbian neurophysiological account of what bridges the gap between what is causally interacted with and what is perceived in the case of physical objects can also provide for the perception of sets. I think this lends considerable credibility to the set theoretic realist's claim that sets are perceivable, but, as mentioned earlier, I don't intend to tie set theoretic realism to this particular neurophysiological theory. Benacerraf-style worries are based on a deep conviction that a certain kind of explanation cannot be given; see Putnam's rhetorical outburst:

> What neural process, after all, could be described as the perception of a mathematical object? (Putnam (1980), 430)

My goal in this section has been to indicate that there is at least one plausible answer to this question.

3. Intuition

So far, I have argued that sets, suitably understood, can be perceived. While this may be enough to answer the Benacerraf-style worries about compromise Platonism, it provides only the barest beginning of an account of set theoretic knowledge. What is the relation, for example, between our knowledge of particular facts about particular sets of physical objects, and our knowledge of the simplest set theoretic axioms? How, for example, do we come to know that any two objects can be collected into a set with exactly those two members, or that the members of any two sets can be combined into a set that is their union? These general beliefs underlie two of the most elementary set theoretic axioms—Pairing and Union—and our epistemology must account for them.

Given our analogy between mathematics and natural science, let's first ask for the source of comparable beliefs in the physical sciences. Rudimentary accounts of elementary physical knowledge most often begin with simple, enumerative induction: every swan in my sample is white; therefore, I conclude, all swans are white. It might be argued that our most basic general set theoretic beliefs are justified in the same way, but this line is unconvincing.[77] Do I test

[77] The idea that mathematics is a simple inductive science goes back to Mill (1843), bk. 2, chs. 5 and 6. Objections have been mounted by many writers, including Frege (1884), §§9–10, Hempel (1945), and Kim (1981).

to see whether or not the two sets of fingers on my right and left hands can be combined to form a larger set of fingers? No, once I am able to understand the question, the answer is obvious. Would I be more sure that two sets can be combined if I successfully combined a wide variety of sets containing different kinds of objects? No,[78] but the observation of white swans in a wide variety of different environments does add support for the claim that they are all white. Evidently, particular observations provide a very different type of support for general hypotheses like 'any two sets can be combined' than they do for general hypotheses like 'all swans are white'.

Does this mean that set theory is dramatically different from physical science, after all? Where there is an analogy, there must be differences as well as similarities, but I think we've not yet reached the point at which our comparison between mathematics and natural science is of no further use. Rather, I think that we've filled in the details incorrectly, that our primitive general set theoretic beliefs actually correspond, not to simple enumerative inductions, but to primitive general beliefs about physical objects that are no more subject to simple inductive support than their set theoretic counterparts.

Consider, for example, the child's hard-won beliefs that physical objects exist independently of the human viewer, that they are independent of their state of motion. Can these be tested by observation of particular examples? Does the observation that a teacup persists when it's moved across the table add support over and above that provided by similar observations of coffee-cups? Evidently not. These are primitive general beliefs about physical objects that are not supported by simple enumerative induction. We cannot check to see whether or not physical objects persist when no one is observing them, but we believe it nevertheless, and beliefs of this sort appear at the most elementary levels of our physical theory of the world.

Hebb's analysis of neural operations provides one possible account of their source. Recall that repeated viewings of a triangle's apex lead to the development of a first-order cell-assembly that

[78] In childhood, such manipulations with a variety of sets helped engender my ability to see sets in the first place, but once I have this ability, my conviction that two sets can be combined doesn't depend on my testing a variety of the sets I can now see.

responds to angles of like magnitude and orientation. Assemblies at this level respond to particular contours in the visual field, simple tastes, localized tactile pressures, and the like. The same mechanisms that produce assemblies at this simple level are capable of producing higher-order assemblies as well, such as the integrated triangle assembly. Similarly, repeated viewing of a single physical object from one perspective produces a second-order assembly which integrates the first-order assemblies for the contours of the object's parts. And finally, manipulation of the object, or seeing it in motion, permits the development of a third-order assembly integrating the assemblies for the object's various perspectives. At this point, perception of the object involves a complex phase sequence of stimulations of all these assemblies.[79]

With these mechanisms in place, the subject is able to perceive an independent, continuing physical object. But there is no reason to suppose that neural developments come to an end here. Hebb suggests:

> Fundamental . . . is the generalized idea of a thing, an object, a space-occupying and sense-stimulating *something*, as the activity of a higher-order cell-assembly made up of neurons that are usually or always active in the relatively small number of different situations of infancy when a visible, tangible object attracts attention. Those neurons must be a small proportion of the total number excited on any one of such occasions but may still be a large number in absolute terms. The theoretical possibility of such an assembly is clear and the psychological support for its existence is also clear. (Hebb (1980), 109)

This fourth-order assembly would correspond to the general concept of a physical object. It would be stimulated during the phase sequence associated with perception of a particular physical object when attention is drawn to its more general features.

A subject with such an assembly would automatically have various general beliefs about the nature of the objects that stimulate it; we might say that these beliefs are 'built into' the cell-assembly much as three-sidedness is built into the triangle-detector in the form of mechanisms stimulating eye movements from one corner to another, or three-angledness in the form of the three first-order corner components. Crudely put, the very structure of one's

[79] See Hebb (1980), 107–8. Hebb cites work on hierarchies of neurons that provides physiological support for the idea of higher-order assemblies.

triangle-detector guarantees that one will believe any triangle to be three-sided. Similarly, anyone with a general physical object assembly would believe that physical objects are 'space-occupying and sense-stimulating', to use Hebb's examples, or observation- and trajectory-independent, to use examples mentioned earlier. These are primitive, very general beliefs about the nature of whatever stimulates the appropriate higher-order assembly. I call them 'intuitive beliefs'.

What goes for physical objects should also go for sets: the development of higher-order cell-assemblies responsive to particular sets gives rise to an even higher-order assembly corresponding to the general concept of set. The structure of this general set assembly is then responsible for various intuitive beliefs about sets, for example that they have number properties, that those number properties don't change when the elements are moved (barring mishap), that they have various subsets, that they can be combined, and so on. And these intuitions underlie the most basic axioms of our scientific theory of sets.[80]

I've suggested that the very structure of one's general physical object assembly gives one some intuitive beliefs about physical objects, for example that objects can look different from different points of view or that they don't cease to be when we cease to see them, and that one's general set assembly gives one intuitive beliefs about sets, for example that any two objects can be collected into a set. Of course, it's deceptive to describe these beliefs in this way, because they are not, in fact, linguistic. A child of two, for example, is perfectly capable of perceiving particular physical objects, and thus has or will soon have intuitive beliefs about physical objects in general, but she lacks the vocabulary to express them as I have. In short, such beliefs are accessible to those who lack the linguistic terms, but not to those who lack the concept.[81] When a term for the concept is introduced, linguistic expressions of intuitive beliefs will naturally seem obvious, too obvious to benefit from simple enumeration.

Granting, then, that there are such primitive, general beliefs about physical objects and about sets, we should inquire into their epistemological status. It's already been noted that they are not

[80] Hebb (1980, esp. pp. 122–4) discusses mathematical cases as part of his theory.

[81] See Hebb (1980), 108–9.

inductively supported, but this should not be taken to imply that they are, in any sense, infallible. The first and most obvious source of potential error is in the uncertain transition from intuitive belief to linguistic formulation. Because all but the most severely disabled eventually attain the requisite cell-assemblies, widespread agreement about an attempted linguistic formulation constitutes one of its best claims to correctness. On this account, then, it is legitimate to suspect the claim of a single scientist as to the intuitiveness of a certain principle if few others share this opinion.[82]

A second source of potential error is the distinct possibility that the intuitive belief itself is false.[83] We might be radically mistaken in the concepts we form; perhaps stimulation by aspects of things causes us to form assemblies which embody features very different from those of the things themselves; I suppose it is possible that physical objects do in fact disappear when no one is watching, or that sets actually don't have subsets, hard as it is for us to imagine such things. Some intuitive beliefs have in fact been falsified by the progress of science, for example the belief that, at any given moment, a physical object is in a certain location and moving at a certain speed, or that every property determines a set of things with that property.[84] Thus, in scientific contexts, intuitive beliefs must be tested like any other hypothesis, and like any other hypothesis, they can be overthrown.

[82] For example, the beliefs often described as Gödel's 'intuitions' about various consequences of the continuum hypothesis (1947/64, pp. 479–80) are not widely shared. I should also note that these judgements of Gödel's are more esoteric than the primitive propositions I've been characterizing as linguistic formulations of intuitive beliefs. In many such cases—when advanced ideas are called 'intuitive'—what is really at work is a hunch or conjecture based on mathematical experience, a theoretical judgement of the sort a natural scientist makes when her familiarity with the field suggests that this, not that, is the sort of theory likely to work. For the set theoretic realist, this counts as a theoretical, rather than an intuitive, justification. I'll come back to Gödel's case briefly in ch. 4, sect.3, below.

[83] Or 'incorrect' in some sense. There is some difficulty with classifying intuitive beliefs as true or false because they are non-linguistic, and probably non-propositional. Still, they can be classified as correct or incorrect, as tending towards success or failure in their behaviour-guiding function. (See Goldman (1977), 276.) I'll equivocate a bit and continue to use the words 'true' and 'false'.

[84] The set theoretic principle that every property determines a set of things that have that property is called 'unlimited comprehension'. It is false because it would allow the formation of Russell's set, which then leads to paradox. It has been replaced in axiomatic set theory by Zermelo's Separation Axiom, which asserts only that every property determines the set of things in a previously given set with that property; within a fixed set, we're allowed to 'separate' the elements with that property from the rest. See Enderton (1977), 4–6, 20–1.

This highlights the central epistemological question: does the intuitiveness of a belief count in its favour? The extent to which a claim strikes us as obvious, and the degree of community agreement on this degree of obviousness, both constitute evidence that the claim is a good linguistic version of a primitive intuitive belief. We've already seen that intuitive beliefs themselves can be false, that intuitiveness is not conclusive evidence, that it can be outweighed by opposing theoretical evidence, but we now ask whether it provides any support at all for the truth of the claim. There is no doubt that such evidence has been counted in favour of various axioms in the history of set theory, but is there any rationale for this practice?

In line with the earlier discussion of Gettier's problem and the causal theory, we might rephrase our question as: does a true intuitive belief count as knowledge? This rendition is deceptive because we've already granted that intuitive support is not enough in itself to fully justify a claim, but I think this formulation can still be used to focus the main problem. Recall that Goldman, in response to Gettier, added a causal requirement to the traditional, justified-true-belief account of knowledge, and that this requirement is seen by many as fatally damaging to Platonism. But, our intuitive beliefs are products of our cell-assemblies, and the processes responsible for generating those—a combination of evolutionary pressures on our ancestors that determine our initial brain formation and the sum of our childhood interactions with physical objects and sets—are causal. They are suitably causal because it is the corresponding general facts about the environment that both exert the evolutionary pressure and provide the childhood interactions. The problem now comes not from Goldman's addition, but from one of the traditional requirements, namely, that the belief be justified: Steve's belief in the Pairing Axiom may be intuitive—the axiom may seem obvious to him and to us—but what if he can offer no justification beyond those feelings?

Turning this question into an outright objection to intuitive evidence would require the assumption that a justification must always take the form of a convincing series of reasons available to the knower. In contemporary epistemology, this is called 'internalism'.[85] The 'externalist', by contrast, insists that a belief can be

[85] This approach goes back to Descartes (1641). For the contemporary debate between internalists and externalists, see Bonjour (1980) and Goldman (1980).

justified even though the knower is ignorant of that justification. Consider, for example, Steve's perceptual belief that there is a tree in front of him. It is generated by a suitable, causal process, but let's suppose that Steve has no knowledge of optics, retinas, or brain function, that he can produce no reasons for his belief. Does this mean that Steve doesn't know there is a tree in front of him? Or, to take a more exotic example from Goldman,[86] consider the case of the professional chicken-sexer. The man looks at a chick and comes to believe that it is male, but he has no awareness of the process by means of which he comes to this judgement. Given that he invariably proves right in his classifications, we are inclined to say he knows the chick to be a male. But he can offer no reasons, no arguments, no explicit justifications.

I side here with the externalist, rejecting the demand that Steve be prepared to justify his belief in the Pairing Axiom. On this view, it is enough that the causal process that generates the belief be 'reliable', that is, the sort of process that generally leads to true beliefs. This is true in the perceptual case, in the chicken-sexing case, and, if our assumptions are correct, in the intuitive case as well. Thus the strength of Steve's conviction that Pairing is obviously true, along with the prevalence of similar convictions in others, supports the claim that Pairing is a good linguistic rendering of an intuitive belief, and the fact that a belief is intuitive lends prima-facie support to the claim that it is true. If Pairing is in fact true, and if further theoretical support is forthcoming—for example, evidence that the axiom is consistent, that it produces theorems of the sort expected, and so on—then it seems Steve's belief can amount to knowledge.

One peculiarity of intuitive evidence should be noted. The acquisition of intuitive beliefs doesn't depend on any particular experience, that is, any sufficiently rich course of experience will produce the required cell-assemblies:

> In the case of visual perception, what assemblies develop and become the basis of perception is fully dependent on an innate property of the organism, the reflex responsiveness of the eye muscles, as well as on the innately determined structure of the striate and peristriate cortex, which—according to the theory—is what makes the formation of cell-assemblies

[86] Goldman (1975), 114. He also defends externalism in Goldman (1976 and 1979).

inevitable, given exposure to a normal visual environment. (Hebb (1980), 105)

So, though experience is needed to form the concepts, once the concepts are in place, no further experience is needed to produce intuitive beliefs. This means that in so far as intuitive beliefs are supported by their being intuitive, that support is what's called 'impurely a priori'. Notice, however, that it doesn't follow that even these primitive mathematical beliefs are a priori. Without the corroboration of suitable theoretical supports, no intuitive belief can count as more than mere conjecture.

I should also emphasize that the particular mathematical intuitions discussed here are by no means the end of the story. Because of my interest in set perception, I've concentrated on the intuitions involved in the perception of small sets of medium-sized physical objects, but I don't want to suggest that these discrete intuitions, the ones that underlie number theory and the beginnings of set theory, are the only intuitions relevant to a complete account of set theoretic knowledge. To begin with, part of perceiving a physical object is perceiving it as existing in external space, that is, perceiving a boundary between it and the space surrounding it. The concept of this space and the ability to perceive a boundary or shape develop along with the concept of physical object itself, as a result of the child's interactions with the environment.[87] As in the case of the triangle-detector, neurological correlates of boundaries are partly constituted by motor stimulations for eye-movements along edges of a figure; perceived shapes are closely related to acts of tracing.

Thereafter, just as children gradually develop the ideas of inclusion, collection, and numeration leading to the discrete set concept, they also develop parallel ideas based on part/whole and enclosure relations rather than inclusion, proximity and distance rather than collection, measurement rather than number.[88] These developments, beginning, as their set theoretic counterparts do, in perception and action, lead eventually to the perception of lines—edges, intersections of planes, trajectories—as continuous structures. By the age of ten or twelve, the child expresses intuitive

[87] See Piaget (1937), or Phillips (1975), chs. 2–4.
[88] See Piaget and Inhelder (1948).

beliefs about geometric figures that reveal a primitive notion of continuity.

These intuitions play a role in our systematic thinking in geometry and analysis that is analogous to the role of intuitions of discrete collections in arithmetic and that of intuitions of physical objects in natural science. In part, for example, they lead to the obviousness of density, and even the Dedekind-style continuity axiom.[89] The ability of set theoretic methods to provide a consistent rendering of our confused intuitive beliefs about the relation of the line to its smallest parts is one of its greatest achievements, and one of its strongest supports.[90]

Finally, let me repeat: I'm not suggesting that all, or even most, epistemic support for our theory of sets is intuitive. In many cases, set theoretic methodology has more in common with the natural scientist's hypothesis formation and testing than with the caricature of the mathematician writing down a few obvious truths and proceeding to draw logical consequences. As the science/mathematics analogy would indicate, our set theoretic hypotheses demand theoretical or extrinsic support, that is, support, as in natural science, in terms of verifiable consequences, lack of disconfirmation, breadth and explanatory power, intertheoretic connections, simplicity, elegance, and so on. A preliminary description of the important role of such non-intuitive, non-demonstrative justifications in modern set theory will be sketched in Chapter 4.

4. Gödelian Platonism

In this chapter, I've sketched the epistemological beginnings of set theoretic realism, a version of compromise Platonism. The differences between this view and Quine/Putnam Platonism should be clear enough from the discussion in the final section of Chapter 1, but its relationship to Gödelian Platonism has been less explicitly

[89] Density is the claim that between any two points there is another. Dedekind's axiom is a bit more complex. If the points on a line are divided into two (non-empty) groups, and the points of one group are all to the left of all points in the other, then there is either a right-most point in one group, or a left-most point in the other. In other words, if you cut a line in two, there's always a point at which you cut it. See Dedekind (1872).

[90] See ch. 3, sect. 1, below.

drawn. I'll conclude this chapter by further detailing my considerable debt to Gödel and by locating our major disagreement.[91]

Gödel's Platonism rests on an analogy between mathematics and natural science, an analogy he traces back to Russell.[92] Mathematical things are taken to be as objective as the objects of the natural sciences: '[Sets] may . . . also be conceived as real objects . . . existing independently of our definitions and constructions' (Gödel (1944), 456). The next stage of the analogy is epistemological: 'The analogy between mathematics and a natural science . . . compares the axioms of logic and mathematics with the laws of nature and logical evidence with sense perception . . .' (Gödel (1944), 449); 'we do have something like a perception also of the objects of set theory . . .' (Gödel (1947/64), 483–4). The problem, for Gödel as for the set theoretic realist, is to explicate this perception-like connection.

To flesh out his proposal, Gödel considers the details of our ordinary perceptual experience and concludes that

> in the case of physical experience, we *form* our ideas . . . of . . . objects on the basis of something else which *is* immediately given. . . . That something besides the sensations actually is immediately given follows . . . from the fact that . . . our ideas referring to physical objects contain constituents qualitatively different from sensations or mere combinations of sensations, e.g., the idea of object itself . . . (Gödel (1947/64), 484)

Here the set theoretic realist agrees, and proposes that the source of this extra constituent, what bridges the gap between retinal stimulation and perception, is the neural cell-assembly. She agrees also that these conceptual elements of perceptual experience 'represent an aspect of objective reality, but, as opposed to the sensations, their presence in us may be due to another kind of relationship between ourselves and reality' (Gödel (1947/64), 484). For the set theoretic realist, this 'other kind of relationship' is the complex causal process that produces the cell-assembly, namely,

[91] There are others. First, as was emphasized in ch. 1, sect. 4, any compromise Platonist will place more weight than Gödel does on the indispensability arguments. Second, as remarked in sect. 3 above, Gödel's 'intuitions' cover more esoteric cases than the set theoretic realist's, cases involving what I would take to be theoretical rather than intuitive evidence. In fact, Gödel's text doesn't explicitly count these as intuitions, so on this point I may be disagreeing more with his readers than with Gödel himself.

[92] See Gödel (1944), 449. For Russell's views, see Russell (1906), (1907), and (1919), esp. p. 169.

the evolutionary pressures on our ancestors and our childhood experiences with objects.

Now what of mathematics? Gödel goes on to suggest that 'Evidently the "given" underlying mathematics is closely related to the abstract elements contained in our empirical ideas' (Gödel (1947/64), 484). Again the set theoretic realist agrees, understanding the 'given' here to mean the intuitive beliefs that underlie the simplest set theoretic axioms. These intuitions are 'closely related' to the cell-assemblies responsible for the 'abstract elements' of our perceptual beliefs: the object and set concepts. In such cases, indeed, 'the axioms force themselves upon us as being true'.[93] This sort of obviousness is evidence for the intuitive basis of the axiom in question.

Finally, Gödel and I agree that not all axioms can be justified on intuitive grounds, that the science/mathematics analogy should be extended one step further, to the level of scientific hypotheses:

> the axioms need not necessarily be evident in themselves, but rather their justification lies (exactly as in physics) in the fact that they make it possible for these 'sense perceptions' to be deduced . . . (Gödel (1944), 449)

> besides mathematical intuition, there exists another (though only probable) criterion of the truth of mathematical axioms, namely their fruitfulness in mathematics and, one may add, possibly also in physics. (Gödel (1947/64), 485).

He even supplies a list of various particular forms these justifications might take:

> Success here means fruitfulness in consequences, in particular in 'verifiable' consequences, i.e., consequences demonstrable without the new axiom, whose proofs with the help of the new axiom, however, are considerably simpler and easier to discover, and make it possible to contract into one proof many different proofs. The axioms for the system of real numbers, rejected by the intuitionists, have in this sense been verified to some extent, owing to the fact that analytical number theory frequently allows one to prove number-theoretical theorems which, in a more cumbersome way, can subsequently be verified by elementary methods. (Gödel (1947/64), 477)

And in a passage quoted earlier:

> There might exist axioms so abundant in their verifiable consequences,

[93] Gödel (1947/64), 484.

shedding so much light upon a whole field, and yielding such powerful methods for solving problems . . . that, no matter whether or not they are [intuitive], they would have to be accepted at least in the same sense as any well-established physical theory. (Gödel (1947/64), 477)

Examples are described in Chapter 4 answering to each of these forms of justification and others like them.

This quick survey of Gödel's writings surely suggests that set theoretic realism should be understood as a further development along the same lines; indeed it was the passage about the relationship between 'the "given" underlying mathematics' and 'the abstract elements contained in our empirical ideas' that set me on this road in the first place. Still, Chihara has rightly pointed out that my quotations and references are highly selective, sometimes deceptively so.[94] In particular, I neglected to site Gödel's assertion that 'the objects of transfinite set theory . . . clearly do not belong to the physical world and even their indirect connection with physical experience is very loose . . .' (Gödel (1947/64), 483). And, in featuring Gödel's claim that 'we do have something like a perception also of the objects of set theory', I omitted the qualifier 'despite their remoteness from sense experience'.[95] Gödel's insistence on the traditional Platonistic characterization of mathematical objects as non-spatio-temporal clearly disqualifies set theoretic realism as a straightforward development of his thinking.

Of course, my motivation for bringing sets into the physical world and for tying mathematical intuition so closely to ordinary perception is naturalism; set theoretic realism seems to me the most promising approach for bringing mathematical ontology and epistemology into line with our overall scientific world-view. Gödel, by contrast, not only characterizes sets as traditional Platonistic entities, he also goes on to postulate a realm of non-physical, non-spatial, mentalistic monads, as well. It is worth asking why.

Consider the following problem: the human mind is finite and

[94] See Chihara (1982), written partly in reply to Maddy (1980). (John Burgess once warned me that I wasn't distinguishing clearly enough between my own ideas and Gödel's, but this wise counsel fell on deaf ears.) In the course of arguing that my view is not Gödel's, Chihara raises two difficulties for set theoretic realism. The first, concerning the status of the continuum hypothesis, will be touched on in ch. 4, sect. 3, below. The second, concerning singletons, was the inspiration for ch. 5, sect. 1.

[95] Both quotations in this sentence comes from Gödel (1947/64), 483–4.

the set theoretic hierarchy is infinite. Presumably any contact between my mind and the iterative hierarchy can involve at most finitely much of the latter structure. But in that case, I might just as well be related to any one of a host of another structures that agree with the standard hierarchy only on the minuscule finite portion I've managed to grasp. According to Benacerraf, Gödel felt that his monads were immune to this finiteness problem, that they could somehow gain unambiguous access to the full hierarchy.[96] Thus his monadology would succeed where naturalism purportedly fails.

The question of how our finite minds make contact with the infinite is given a very general formulation by Kripke in his interpretation of Ludwig Wittgenstein.[97] In its strongest form, this argument, the so-called 'rule-following argument', applies not only to the case of the mathematical infinite, but also to any rule with an indefinite number of potential applications. Consider, for example, our ordinary use of the word 'triangle'. Kripke's Wittgenstein[98] argues that all our training, our past usages, our mental images, our stated and unstated intentions, our associations, and so on, cannot predetermine whether it is correct or incorrect to call a given figure a triangle. This is because, for example, these constraints can all be perversely interpreted to conform to the set of things that are triangular up to now or square after now as completely as they do to the set of triangular things. So nothing we've associated with the word 'triangle' can predetermine whether 'triangle' should now be applied to this figure △ rather than that □. The argument doesn't depend on the fact that there are indefinitely many triangles; it can also be applied to show that nothing predetermines whether this person or that is correctly referred to as 'Kripke', because everything I've associated with the name 'Kripke' can now be interpreted as applying as well to a being consisting of Kripke's space-time worm up till now and Putnam's afterwards.

These conclusions may seem outrageous, and indeed they should, but I will leave the interested reader to more complete expositions

[96] Benacerraf made this observation in an oral presentation that became Benacerraf (1985). As far as I know, the details of Gödel's monadology and how it is to overcome this problem are as scant as his account of intuition.

[97] See Kripke (1982). For Wittgenstein's views, see Wittgenstein (1953) and the posthumous (1978).

[98] I put it this way because there is some debate over whether or not Kripke's argument is really Wittgenstein's.

of this fascinating paradox. All I want to do here is to indicate—in light of Gödel's motivations—what I take to be the main ingredients of a naturalistic solution.[99] It begins, of course, with the idea that my linguistic training sets up a neural connection between the word 'triangle' and my triangle-detector. This is not a complete reply, however, because of the inevitable physical limitations on my neural processors. The range of the term 'triangle' cannot be identified with the set of things that stimulate my detector because my detector is sometimes wrong, and even if it weren't, it couldn't respond to triangles too small, too large, or too far away.

At this point, the realist must appeal to the objective fact that triangles are more like one another than like squares, that is, to the fact that there is an objective difference, noted in the second section of Chapter 1, between random and natural collections.[100] Our best theory of what my detector responds to involves not the sceptic's random collections, but those collections science takes to be the natural ones. Triangles, then, are those things belonging to the natural collection that includes most of the things that stimulate my detector. This distinction between natural and random may be difficult to pin down; it may be a matter of degree rather than all or nothing, it may ultimately require an ontological commitment to properties or universals, but it is a crucial part of the naturalist's world view, and it serves to rule out the gerrymandered interpretations on which the sceptic's arguments depend.[101]

I have tried in this section to clarify the nature and the origin of my central disagreement with Gödel. This should not be allowed to obscure my obvious debt to his thinking, both in the formulation of mathematical realism in Chapter 1 above and in the account of mathematical perception and intuition offered in this chapter. His influence will again be prominent in Chapter 4, when I take up the subject of theoretical evidence and justification, but before turning to that more esoteric epistemological study, I want to touch on another Benacerrafian worry about Platonism, an ontological one this time.

[99] For a fuller account of both the paradox and its naturalistic solution, see Maddy (1984*b*).

[100] And between natural and unnatural individuals, to deal with the 'Kripke' case.

[101] Putnam's (1977) and (1980) present a puzzle in some ways similar to Kripke's version of the rule-following argument. In reply to Putnam, Merrill (1980) and Lewis (1983; 1984) use natural kinds much as Maddy (1984b) does in reply to Wittgenstein.

3

NUMBERS

1. What numbers could not be

The widely cited epistemological challenge to Platonism has been my focus so far, and I will return to epistemology in the next chapter to take up an equally important, though less talked-of aspect of mathematical knowledge. But before doing that, I would be negligent not to discuss a celebrated ontological challenge to Platonism which some observers consider nearly as daunting as the epistemological.[1] The set theoretic realist's answer is implicit in what has already been said, but it will take some effort to see this and to draw out the details.

Let me frame the problem with another touch of simplified history. The emergence of set theory as a foundational theory can be traced to two separate lines of development that eventually converged.[2] The first of these began with concern over the foundations of the calculus.[3] The much- and deservedly-criticized infinitesimals were replaced by Weierstrass's theory of limits, which in turn depended on a theory of the real numbers proposed in different forms by Cantor and Dedekind.[4] These accounts are thoroughly set theoretic; Dedekind's, for example, makes essential use of infinite sets of rational numbers. By these means, he clarifies the notion of continuity, defines the real numbers, and proves that they are, in fact, continuous. Thus, by identifying real numbers with certain sets (called 'Dedekind cuts'), Dedekind obtains a rich

[1] Among those bothered are Benacerraf (1965), Jubien (1977), Kitcher (1978; 1983, p. 104), Field (1980, p. 126, n. 66; 1989, pp. 20–5), and Resnik (1981, p. 529).

[2] I'll concentrate here on two foundational motivations. Cantor's more purely mathematical interests will be discussed in the next chapter.

[3] This was touched on in ch. 1, sect. 4, above. The resulting worry about the infinite produced the three great schools of thought in the philosophy of mathematics that flourished in the early decades of our century.

[4] Cantor's theory appears in Cantor (1872). See Dauben (1979), ch. 2, for discussion. Dedekind's account, developed somewhat earlier, first appeared in Dedekind (1872). Enderton (1977), ch. 5, presents Dedekind's more elegant formulation.

and explanatory theory of their nature and behaviour that puts the calculus and higher analysis on a consistent foundation. To this day, set theory provides our best account of mathematical analysis, which in turn plays a central role in our most successful physical theories. This achievement is one of the strongest theoretical supports for the mathematical theory of sets; it plays an indispensable role in our best theory of the world.

Meanwhile, from a more philosophical orientation, Frege was concerned to provide a foundation for ordinary arithmetic.[5] He was scandalized by the lack of understanding, even among mathematicians, of the fundamental concepts of their subject, in particular, the concept of a natural number. Frege's aim was to show that arithmetic is in fact a branch of logic, but in doing so he made use of extensions of concepts, that is, often infinite collections. This project failed, as has been noted,[6] but its core has been incorporated into modern set theory. As with the reals, the natural numbers are identified with certain sets, and all their basic properties, once assumed as axioms, become provable, explicable.[7] Thus modern set theory also provides our best account of the natural numbers, another key ingredient in our overall theorizing.

Here we have a scientific success story. It depends, however, on identifying numbers with sets. In the opening pages of Chapter 1, I suggested that the philosophy and foundations of set theory are of interest regardless of whether or not all mathematical objects are properly taken to be sets, but in this case, where the theoretical justification of the subject, indeed the very motivation for the subject, depends so centrally on these particular identifications, philosophical questions about the nature and propriety of set theoretic reduction cannot be put aside. It is this question—of the relationship between sets and numbers—that I want to consider in this chapter.

[5] See Frege (1884), introd.

[6] See ch. 1, sect. 4, above.

[7] I don't mean to suggest that all proofs are also explanations, but some clearly are. For example, the set theoretic account of natural numbers tells us why multiplication is commutative: because the cross product $A \times B$ is equinumerous with $B \times A$. (See Drake (1974), 52.) This same fact can be proved from the Peano axioms, but the proof requires a series of clever lemmas and sheds little light on why the theorem is true. (See Enderton (1977), 81–2.) Steiner (1978) makes an effort to characterize mathematical explanation.

To see the source of the problem, let's return for the moment to Frege and reconsider his central ontological query: what is a number?

Take a numerical statement like 'there are two boys playing in the garden'. I've already reviewed[8] Frege's observation that the subject of this numerical ascription cannot be the mere physical stuff that makes up the boys, because that physical stuff could be divided into units in various different ways—two boys, twenty boy-parts (heads, torsos, arms, hands, legs, and feet), millions of cells, even more molecules, still more atoms, and so on. What has changed, Frege argues, from the occasion when we judge that the stuff is two boys to that when we judge it to be millions of atoms is the substitution of the concept 'atom in this mass of physical stuff' for the concept 'boy in this mass of physical stuff'. This leads him to the conclusion that 'the content of a statement of number is an assertion about a concept'.[9]

But to identify the content of a statement of number is not yet to identify the number itself. Frege's analysis suggests that number is something various concepts can share, for example 'boy in the garden' and 'hand on my keyboard'. In Fregean terms, this is to say that two is a concept under which other concepts can fall, a second-order concept, as it were. But, however straightforward this might sound, Frege insists, for reasons I'll consider later, that a number is a thing, a 'self-subsistent object',[10] rather than a concept. So, instead of saying that two is the concept that applies to any concept equinumerous[11] with 'boy in the garden', he identifies the number with the extension of that concept, that is, with the collection of all concepts equinumerous with 'boy in the garden'.

The trouble with this account is that two turns out to be a very large collection indeed. In a set theory like Frege's, with the (false) principle of unlimited comprehension, we could form this set, but alas, we could also form the paradoxical Russell set. When this inconsistent naïve set theory is replaced by the contemporary

[8] In ch. 2, sect. 2, above.

[9] Frege (1884), § 46.

[10] Frege (1884), § 55.

[11] This sounds circular, defining the number two in terms of equality in number, but in fact, Frege defines 'equinumerous' without reference to numbers. Two concepts are equinumerous if there is a one-to-one correspondence between the things falling under the first and the things falling under the second. See Frege (1884), § 63, or Enderton (1977), 128–9.

axiomatic version based on the iterative conception, two, as defined here, no longer exists. To see this, notice that new two-membered sets are formed at each stage, so there is no stage at which the set of all two-membered sets is formed.

Thus, in contemporary set theory, two is customarily identified not with the (non-existent) collection of all two-membered sets, but with some particularly convenient example of a two-membered set. Standardly, the natural numbers are taken to be the set of finite von Neumann ordinals, that is, $\emptyset$ for 0, $\{\emptyset\}$ for 1, $\{\emptyset, \{\emptyset\}\}$ for two, $\{\emptyset, \{\emptyset\}, \{\emptyset, \{\emptyset\}\}\}$ for 3, and so on.[12] Other, similar, identifications would do as well, for example Zermelo's $\emptyset$ for 0, $\{\emptyset\}$ for 1, $\{\{\emptyset\}\}$ for 2, $\{\{\{\emptyset\}\}\}$ for 3, and so on.[13] And it is this fact that generates an ontological question about numbers.

Once again, the version of this problem most exhaustively cited and discussed in the contemporary philosophical literature derives from a paper by Paul Benacerraf, this time 'What numbers could not be'.[14] Benacerraf asks us to consider the education of two hypothetical young children, Ernie (for Ernst Zermelo) and Johnny (for John von Neumann).

Ernie and Johnny are both brought up on set theory. When the time comes to learn arithmetic, Ernie is told, to his delight, that he already knows about the numbers; they are $\emptyset$ (called 'zero'), $\{\emptyset\}$ (called 'one'), $\{\emptyset, \{\emptyset\}\}$ (called 'two'), and so on. His teachers define the operations of addition and multiplication on these sets, and when all the relabelling is done, Ernie counts and does arithmetic just like his schoolmates. Johnny's story is exactly the same, except that he is told that the Zermelo ordinals are the numbers. He also counts and does arithmetic in agreement with his schoolmates, and with Ernie. The boys enjoy doing sums together, learning about primes, searching for perfect numbers, and so on.

But Ernie and Johnny are curious little boys; they want to know everything they can about these wonderful things, the numbers. In

[12] See e.g. Enderton (1977), 68. Von Neumann's proposal is contained in his (1923), 347.

[13] See Zermelo (1908*b*), 205. In fact, there are reasons why Zermelo's version isn't as good as von Neumann's. For example, von Neumann's account works just as well for infinite numbers as for finite, and its 'less than' relation is extremely simple: membership.

[14] Benacerraf (1965). See also Parsons (1965). The argument is further developed by Kitcher (1978).

the process, Ernie discovers the surprising fact that one is a member of three. In fact, he generalizes, if n is bigger than m, then m is a member of n. Filled with enthusiasm, he brings this fact to the attention of his favourite playmate. But here, sadly, the budding mathematical collaboration breaks down. Johnny not only fails to share Ernie's enthusiasm, he declares the prized theorem to be outright false! He won't even admit that three has three members!

According to Benacerraf, the moral of this sad story—or one of them—goes like this: if numbers are sets, then they must be some particular sets. Any choice of particular sets will exhibit properties that go beyond what ordinary arithmetic tells us about the numbers. (Ordinary arithmetic is mute on the subjects Ernie and Johnny debate. Their classmates are puzzled by the very questions these boys take so much to heart.) If one of these particular choices is the correct one, that is, if one sequence of sets really is the numbers, then there ought to be arguments that tell us which sequence that is.[15] (This doesn't seem to be the sort of question that requires some further, deep number theoretic theorem.) But there are no such arguments. Therefore, numbers are not sets.

Friends of numbers might be prepared to fall back on the position that while numbers aren't sets, still they are objects of some other kind. Suppose, then, that we've identified some sequence of objects suitable for counting and arithmetic, and we claim that the fourth of these (we started from zero) is the number three. Benacerraf argues that this object plays the role of three in our sequence by virtue of its relations to the other members of the sequence, but if it is to be singled out, independently of the sequence, as this object or that, it must have some additional properties. And, he continues, these additional properties will be superfluous to the object's numerical functioning, in the same sense that the properties Ernie and Johnny debated were superfluous, which leads to the question: why should three have these superfluous properties and not some others? If this object really is three, there should be arguments to show that these superfluous properties are the correct ones, but there are no such arguments. Therefore, numbers are not objects at all.

[15] The metaphysician wants more here than the previously remarked arguments from convenience. The fact that von Neumann ordinals are more convenient than the Zermelo ordinals, and hence standard in contemporary set theory, doesn't give us any further reason to think that they *really are* the numbers.

This second conclusion is less firmly supported than the first; the second argument leaves room for the position that numbers are the sort of objects whose non-superfluous properties are all the properties they have.[16] If these non-sets are connected closely enough with sets, perhaps even the explanatory virtues of the set theoretic reduction can be preserved without the actual identification of numbers with sets. Thus, for example, Cantor suggests that natural numbers are separate entities 'abstracted' from equinumerous sets, and Dedekind that the reals are 'associated' with the corresponding cuts. This sort of move obviously flaunts ontological economy—it's inefficient to overburden our theory with more things than we need to make it work effectively—but worse than that, it requires an account of the sort of 'abstraction' or 'association' involved. Neither Cantor nor Dedekind provide this.[17]

Of course, if we take Benacerraf's argument that natural numbers aren't sets to be persuasive, as I think we should, an analogous line of thought shows that the reals can't be sets either. We could tell a story of Georgie (for Georg Cantor) and Rich (for Richard Dedekind), one of whom learns that the reals are Dedekind cuts and the other of whom that they are Cantor's fundamental sequences.[18] The rest of the story follows as before, and the conclusion: real numbers aren't sets.[19]

If numbers aren't sets after all, the story of scientific success recited at the beginning of this section is called into question. If its illumination of the theories of natural and real numbers is to count as evidence for the theory of sets, we need to understand the true nature of the ontological relationship between numbers and sets. If not identity, then what?

2. Numbers as properties

Assuming that numbers aren't sets, the set theoretic realist faces the

[16] Steiner (1975*a*), 88–92, suggests a move of this sort in his reply to Benacerraf, but only in an epistemological sense: 'we accept mathematical objects, contra Benacerraf, but we agree that the only things to know about these objects of any value are their relationships with other things' (p. 134).

[17] For a scorching attack on Cantor's notion of abstraction, see Frege (1979), 68–71.

[18] These depend on an idea that goes back to Cauchy. See Enderton (1977), 112.

[19] The same goes for other set theoretic reductions. See e.g. Kitcher (1978) on ordered pairs. The problem for ordered pairs carries over to functions, understood as sets of ordered pairs, and so on.

prospect of adding a new type of entity to her ontology and an extra epicycle to her epistemology. And, to preserve the explanatory force of the standard set theoretic reduction, she must also describe a relationship between sets and numbers that makes the behaviour of, for example, the von Neumann ordinals somehow relevant to our understanding of numbers. Finally, the efforts to naturalize the epistemology of set theory will be wasted if the account of numbers isn't also naturalistic.

Let's begin then with the suggestion that the epistemology for numbers should be as similar as possible to that given for sets. In that case, numbers must also be located in space-time. Where, then, is the number ten?

The easy answer is: ten is located where the set of my fingers is located, in motion over the keys of my word processor. But if this is right, then ten is also located where the starting line-up of any American League baseball team is located, and on the *Times* best-seller list, and many other places. Now a set of physical objects can have a discontinuous location—the set of Angel starting baseball players is located in left field, right field, second base, and even in the dugout, with the designated hitter—but only part of the set is located in each of these places, while the number ten is fully present in each and every ten-element set. By traditional criteria, this makes the set of baseball players a particular and the number ten a universal.[20]

In terms of the science/mathematics analogy, then, the idea is this: set theory is the study of sets and their properties, of which number is one, just as physics is the study of physical objects and their properties, of which length (for example) is one. To see how this works in a bit more detail, consider the formal features of the quantity 'length'.[21]

First, objects with a given physical quantity, like length, are comparable; they form a linear ordering with respect to that quantity. For example, there is a simple linear ordering of medium-sized, easily movable objects that goes like this: A is shorter than B if one end of B extends beyond the end of A when they are laid side by side with the other ends coincident. A method that also works for stationary objects might go like this: A is shorter than B if a string that lies straight with one end at each end of A won't reach both ends of B. Obviously, more sophisticated tests will extend the

[20] See ch. 1, sect. 2, above.
[21] Here I follow Ellis (1966).

linear ordering further. At this point, a scale of measurement can be assigned, as long as it agrees with the established linear ordering, that is, as long as the ordering of the numerical assignments agrees with the linear ordering. This is not difficult: comparison with any fair ruler measures length in yards. And finally, the same quantity can be measured on different scales; length is detected by metre sticks as well.

Now compare the case of number properties. Sets of easily movable objects are directly comparable with respect to number; A is less numerous than B if each member of A can be set beside a unique member of B in such a way that there are elements of B left over. The linear ordering of simple sets this produces can be extended to all sets using the mathematical idea of a one-to-one correspondence: A is less numerous that B if there is a one-to-one correspondence between the members of A and a proper subset of the members of B. Scale can be assigned, for example, by comparison with the English number words, that is, by counting in English.[22] Or, more elaborately, in terms of one-to-one correspondence with the set of von Neumann ordinals.

The only disanalogy is that there is no room for measuring sets on different scales, because there is no room for an arbitrary choice of unit: sets come with their elements already individuated. This is how sets avoid Frege's objection to physical masses as bearers of number properties. While it is arbitrary to assign two to the physical mass that makes up the boys playing in the yard, there is no arbitrariness in assigning two to the set of boys playing in the yard.[23]

This suggestion—that numbers are properties of sets, analogous to physical properties, and in particular, to physical quantities—meets our epistemological desiderata with admirable economy; the naturalistic epistemology previously described for set theory needs no elaboration. Just as the perception of physical objects includes the perception of their properties, so the perception of sets does the same. Indeed, our account of set perception developed from the observation that we perceive their number properties; all that is new here is the further claim that these number properties are the numbers.

[22] Compare Benacerraf (1965), 292: 'The central idea is that [the sequence of number words] is a sort of yardstick which we use to measure sets.'

[23] Yourgrau (1985) disagrees, but for a cogent reply, see Menzel (1988).

Knowledge of numbers is knowledge of sets, because numbers are properties of sets. Conversely, knowledge of sets presupposes knowledge of number; for example, Piaget's studies indicate that subset relations cannot be properly perceived before number properties.[24] From this perspective, arithmetic is part, perhaps the most important part, of the theory of hereditarily finite sets.[25] Neither arithmetic nor this finite set theory enjoys epistemological priority; the two theories arise together. Arguments that arithmetic should not be reduced to set theory because set theory is less certain than arithmetic miss the fact that it makes little sense to separate the epistemological basis of arithmetic from that of finite set theory.[26] Higher set theory is admittedly less certain than the theory of hereditarily finite sets, but this is irrelevant.

Furthermore, the significance of the set theoretic reductions is now clear. The von Neumann ordinals are nothing more than a measuring rod against which sets are compared for numerical size. We learn about numbers by learning about the von Neumann ordinals because they form a canonical sequence that exemplifies the properties that numbers are. The choice between the von Neumann and the Zermelo ordinals is no more than the choice between two different rulers that both measure in metres. The debate between Ernie and Johnny is like an argument over whether an inch is wooden or metal.

Some version of the identification of the natural numbers with properties of sets is considered by both Frege and Benacerraf, and both writers ultimately reject positions of this sort. Benacerraf's conclusion is drawn with little strong conviction—he says only that '"seventeen" *need* not be considered a predicate of [sets]'[27]—but

[24] See Piaget and Szemińska (1941), ch. 7.

[25] A set is hereditarily finite if it is finite, and its members are finite, and the members of its members are finite, and so on. (See Enderton (1977), 256.) It might seem that arithmetic only treats finite sets of physical objects, but the most natural way of understanding such examples as 'here are three pairs of shoes' involves a set of three two-membered sets. The restriction to hereditarily finite sets, rather than to finite sets *simpliciter*, rules out such things as the singleton containing the set of all von Neumann ordinals, hardly the sort of thing covered by ordinary arithmetic.

[26] Notice, for example, that recursion theory can be developed with equal naturalness from numbers or from hereditarily finite sets. An argument of the sort considered in the text appears in Steiner (1975*a*), ch. 2. Sentiments related to my own are voiced by Parsons (1965), 173.

[27] Benacerraf (1965), 284.

Frege's is harder to evaluate. I will briefly consider a few of their reasons.

The most conspicuously cited grounds for their opposition to the property view are grammatical: comparison of the role of number words with ordinary adjectives and predicates,[28] the use of number words with the definite article, their immunity to pluralization,[29] and so on. Of course, grammar is no infallible guide to the actual structure of the world, so such evidence must be taken with a grain of salt. Frege admits as much when he dismisses the contrary grammatical evidence:

> our concern here is to arrive at a concept of number usable for the purposes of science; we should not, therefore, be deterred by the fact that in the language of everyday life number appears also in attributive constructions. (Frege (1884), § 57)

Benacerraf also ignores strong grammatical evidence, the evidence that number words are names, when he later denies that numbers are objects.[30]

In fact, Frege's commitment to the objecthood of numbers seems to waver at the very moment when he presents his own definition. He writes:[31] 'the number which belongs to the concept F is the extension of the concept "[equinumerous with] the concept F"'. To the word 'extension' he appends a footnote: 'I believe that for "extension of the concept" we could write simply "concept".' This surely sounds like a suggestion that numbers could be concepts rather than objects. Frege immediately rejects his proposal: 'But this would be open to the two objections . . .'. The first is the inconclusive grammatical considerations. The second, considerably more interesting, is 'that concepts can have identical extensions without themselves coinciding'. This suggests the fact that concepts are intensional rather than extensional. But whatever the force of

[28] Benacerraf (1965), 282–4; Frege (1884), § 57.

[29] Frege (1884), § 38.

[30] In defence of his own view, Frege admits that it leads to some unusual ways of speaking—e.g. that a number is 'wider or less wide than the extension of some other concept'—but insists that there is nothing 'to prevent us speaking in this way' (1884, § 69). Benacerraf also speaks somewhat oddly when he says that '*any* object [including Laurence Olivier] can *play the role of* 3' (1965, p. 291). Unless it is argued that these oddities are semantic rather than grammatical, a distinction notoriously hard to draw, these are further examples of both writers' willingness to ignore grammatical evidence when need be.

[31] All the quotations in this paragraph come from Frege (1884), § 68.

this objection, Frege does not regard it as decisive. He concludes: 'I am, as it happens, convinced that both these objections can be met; but to do this would take us too far afield for present purposes.' Thus the proposal is rejected on grounds of convenience rather than principle.

It may seem obvious that Frege is leaving open the possibility that numbers are concepts, and depending on the relationship between concepts and properties, perhaps the possibility of a property view as well, but reading Frege is never a simple matter. He holds, for example, that 'the concept horse' must refer to an object, rather than a concept,[32] so it can be argued that this tantalizing footnote suggests no more than that 'the concept "equinumerous with F"' actually refers to the extension of the concept 'equinumerous with F'. On this reading, 'concept' could be substituted for 'extension of the concept' in his original definition because the two actually refer to the same thing, which isn't, by the way, a concept at all. And the reason the concepts 'human being' and 'featherless biped' aren't identical isn't that these expressions have different meanings or stand for different properties, but that identity, and difference, are relations between objects, not between concepts.[33]

I have nothing to contribute to the debate over what Frege actually had in mind here, but leaving Frege himself behind, I'd like to examine the idea that numbers might be concepts and the extent to which the intensionality of concepts stands in the way of such a view. To see what's at stake here, consider a simple arithmetical identity: 2 = S(S(0)).[34] If Frege's definition had read 'concept' in place of 'extension of concept', 2 would be the concept 'equinumerous with the concept "identical with 0 or 1"', and the successor of the successor of 0 would be:

> the concept 'equinumerous with the concept "member of the series of natural numbers ending with S(0)"'

which is

> the concept 'equinumerous with the concept "member of the series of natural numbers ending with the concept 'equinumerous with the series of natural numbers ending with 0'"'.

[32] See Frege (1892*a*), 45.

[33] This line of interpretation appears in Resnik (1965). For other relevant discussions of Frege, see Hodes (1984) and Luce (1988).

[34] 'S' here means 'the successor of'.

Now there's no doubt that the extension of 2, so defined, is the same as the extension of S(S(0)), so defined—each involves being equinumerous with a concept under which two things fall—but if coextensive concepts can nevertheless differ, our simple arithmetical identity is in jeopardy. This is a clear difficulty for the concept view.

Now let's see how this works out on the property view proposed here. If 2 is the number property of, for example, the von Neumann ordinal $\{\emptyset, \{\emptyset\}\}$, then to have the property 2 is to be equinumerous with this set. When successor is defined for von Neumann ordinals, S(S(0)) turns out to be the same set as $\{\emptyset, \{\emptyset\}\}$ itself, so being equinumerous with either one is the same as being equinumerous with the other.[35] The same goes for Zermelo's version. But unless we're able to affirm, for example, the identity

the property 'equinumerous with $\{\emptyset, \{\emptyset\}, \{\emptyset, \{\emptyset\}\}\}$'
= ?
the property 'equinumerous with $\{\emptyset, \{\emptyset\} \{\{\emptyset\}\}\}$'

a new version of the old Benacerraf problem will arise: which of these is 3? Which properties really are the numbers? Those defined in terms of equinumerosity with particular von Neumann ordinals or those defined in terms of equinumerosity with initial segments of the Zermelo ordinals? This is the analogous problem for the property view.

Now let's compare the two problems. A Fregean concept is closely connected to a predicate; indeed, it is the referent of a predicate. Leaving Frege's views on identity aside, let us ask when two predicates are the same. We might insist this is only so when they are typographically identical, but a more flexible notion allows for trivial grammatical transformations, for example, that 'is Sam's only friend' is the same predicate as 'is the only friend of Sam'. On the other hand, anyone would admit that 'is a featherless biped' is a different predicate from 'is human'. This leaves intermediate cases like 'is a bachelor' and 'is an unmarried male'. If we think these two phrases mean the same thing, because 'bachelor' and 'unmarried male' mean the same thing, then a natural account of identity between predicates equates it with synonymy.

[35] The successor of x is $x \cup \{x\}$.

Now what of properties? We might simply specify that two properties are the same just when the predicates that pick them out are synonymous, but we've already seen[36] that this approach doesn't square with our ordinary scientific thinking. Recall that the same length property can be measured on various yardsticks; the property 'measures three inches on this yardstick' is the same as 'measures three inches on that yardstick'. Furthermore, length can be measured in metres as well as yards, so the same property can be expressed by a predicate mentioning yards and another mentioning metres. Yet I think no one would suggest that all these predicates are synonymous. For more dramatic examples, we can turn to ordinary scientific identities like that of 'temperature' with 'mean molecular motion'. These could hardly rest on sameness of meaning.

For scientific properties, then, synonymy is too strong a condition. Somewhere between predicates—individuated by sameness of meaning—and sets—individuated by sameness of membership—there is an intermediate category of properties.[37] One suggestion current among philosophers of science is that the appropriate mode of individuation might be specifiable in terms, not of coextensiveness, but of law-like coextensiveness.[38] Thus 'temperature' = 'mean molecular motion' follows from the laws of physics. By contrast, 'featherless biped' = 'human' is true by the accident of what species happen to exist in our world; no physical law would be violated if there were bipedal fish.

If something along these lines can be made to work for physical properties, our analogy suggests a similar course for mathematical properties, and in particular, for number. Consider again the predicates 'equinumerous with $\{\emptyset, \{\emptyset\}, \{\emptyset, \{\emptyset\}\}\}$' and 'equinumerous with $\{\emptyset, \{\emptyset\}, \{\{\emptyset\}\}\}$'. They aren't synonymous, but they are coextensive, so they are different predicates that determine the same set. But our real concern is whether they express the same

[36] In ch. 1, sect. 2, above.

[37] In the course of his attack on Carnap's distinction between analytic and synthetic—see ch. 1, sect. 4, above—Quine casts serious doubt on the notion of synonymy as well. I don't mean to differ with Quine here. My point is simply that scientific properties aren't individuated by synonymy even if concepts are. If synonymy is a bankrupt concept, so much the worse for concepts; my concern is with properties.

[38] See Putnam (1970), 321.

scientific property; for that they must be coextensive by law rather than by accident. Would a law of mathematics be violated if these two failed to be coextensive? Of course! Their coextensiveness is provable from the axioms of set theory. Thus the neo-Benacerrafian dilemma for property theory—which properties are really the numbers?—dissolves; understood as scientific properties rather than sets or predicates, the von Neumann-style numbers and the Zermelo-style numbers are in fact identical.

I think the part of this story aimed most directly at the problem of multiple reductions for arithmetic carries over to the analogous problem for the real numbers. That is: what makes one set theoretic version of the reals preferable to the others? Answer: nothing; each version serves to detect and measure the same underlying properties. But when it comes to the question that inspired Benacerraf's discussion—what are the natural numbers? or in this case, what are the real numbers?—the answer is, perhaps not surprisingly, a bit more complex.

In fact, I think 'what are the real numbers?' is not as directly analogous to 'what are the natural numbers?' as it at first seems. To see this, compare Dedekind's project with Frege's. Frege was faced with a firmly entrenched linguistic practice which strongly favoured the view that number words were names and numbers were objects, and his job was to clarify the nature of those objects. In Dedekind's case, on the other hand, there was no pre-existing systematic use of real numbers; that was the problem! What pre-existed in this case was the intuitive, geometric line, and Dedekind's job was to produce a system of numbers that would mimic its properties, particularly continuity. So the question Dedekind faced was not 'what are the real numbers?', analogous to Frege's 'what are the natural numbers?', but rather, 'what is continuity?'

The answer Dedekind gave was: continuity is what Dedekind cuts have. And, as the Benacerraf-style argument points out, so do Cantor's fundamental sequences, and other set theoretic versions of the reals. So there is after all a single underlying property that all set theoretic versions of the reals serve to detect, a single property shared by all the particular disparate phenomena they are used to measure, namely continuity. Thus, if there is a proper answer to 'what are the reals?', an answer that runs parallel to our answer to 'what are the naturals?', that is, parallel to 'that which the various set theoretic versions of the naturals serve to detect', then that proper answer is: the real numbers are the property of continuity.

But this sounds odd, and I think the reason it does is clear from what's already been said. The reals, unlike the naturals, are not what needs to be accounted for. Continuity is what needs explication, and the various set theoretic versions of the reals do that. But our intuitive ontology is not the reals and the various set theoretic versions, as it is the naturals and their various set theoretic versions; rather we have the phenomenon of continuity and we have the set theoretic reals, which all explicate that property by exemplifying it. But there aren't any pre-theoretic reals to be identified with anything.

Of all the many mathematical things whose relationship to sets might be questioned, I've considered only numbers, natural and real, because they are most intimately connected with the foundations and justification of set theory itself. For set theoretic purposes, the only other essential is the notion of function, for which I propose a similar treatment. A function is a relation between sets, and a relation is just a two-placed version of a property.[39] Just as von Neumann ordinals give us a way of detecting the number property of a set—we ask whether or not the set is equinumerous with the appropriate ordinal—the set theoretic version of a function should give us a way of detecting whether or not two given sets stand in the appropriate relation of argument to value.

In fact, we do this by identifying the function with a set of ordered pairs.[40] Given two sets, x and y, we can tell if they stand in the functional relation by asking whether, in our chosen set of pairs, x is the first member of some ordered pair of which y is the second member. Naturally, we could do the same job with many other particular sets—e.g. a set of pairs of pairs (e.g. with $\langle\langle\varnothing, x\rangle, \langle\{\varnothing\}, y\rangle\rangle$ in place of $\langle x, y\rangle$) or a closely related set of ordered triples (e.g. with $\langle\varnothing, y, x\rangle$ in place of $\langle x, y\rangle$)—and similarly, we could use various substitutes for the standard set theoretic version of the

[39] This is an oversimplification because the mathematical notion of a function shifted over the centuries from a rule-like relation to an arbitrary mapping. (See ch. 4.) Still, even the most general mapping can be detected by a set, as described in the text.

[40] See Enderton (1977), ch. 3.

ordered pair.[41] All that matters, here as with numbers, is that the set theoretic version give us a convenient way of detecting the relation, the function, that we're interested in.

I've argued that numbers are properties of sets, that elementary arithmetic is the study of the number properties of hereditarily finite sets, that our knowledge of arithmetical facts is of a piece with our knowledge of these finite sets,[42] and suggested similar accounts for real numbers and for functions. This leaves open the metaphysical question of whether or not properties (and relations) should be included as a separate category in the set theoretic realist's ontology. And this, obviously, is just a special case of the age-old debate over universals.[43]

Recall that in our naturalized context, this question reduces to that of whether or not our best overall theory of the world requires us to speak of properties in addition to ordinary physical objects (common-sense realism), various unobservables (scientific realism), and sets (set theoretic realism). The question is just as pressing for physical properties as it is for mathematical ones: should physical properties (like being gold) or physical quantities (like being an inch long) be included, along with physical objects, in the ontology of the natural sciences? Putnam, for one, says yes. For example, he argues, a scientist may conjecture that 'there is a single property, not yet discovered, which is responsible for such-and-such',[44] a statement for which Putnam sees no property-free translation. Nominalistic philosophers of science may either doubt the centrality of these locutions, or disagree about the prospects for translation. Lewis, for example, suggests that the nominalist might treat indispensable statements about properties as asserting the existence of what I've called natural collections or kinds. The idea of naturalness might be taken as primitive, or it might be parsed in terms of various objective similarities between things, without appeal to universals.[45]

[41] The standard version nowadays is the Kuratowski ordered pair (x, y) which is just the set $\{\{x\}, \{x, y\}\}$, but there are many other possibilities. See Enderton (1977), 35–8.

[42] I don't mean to suggest that we can't count hereditarily infinite sets, but again, I deny that this is part of the elementary arithmetic Frege hoped to reduce to pure logic.

[43] See ch. 1, sect. 2, above.

[44] Putnam (1970), 316.

[45] Lewis doesn't advocate either of these views, but he does argue that they are live possibilities. See Lewis (1983), 347–8.

Without pretending to resolve this issue, let me consider an analogous question for our chosen branch of mathematics: should natural numbers, as well as sets, be included in the ontology of the theory of sets? This question can be taken in two senses. First, if we're asking whether there is anything of mathematical significance that can't be said without explicit reference to number properties, then I think the answer is no. This is exactly the moral of the set theoretic reductions: everything we wanted out of numbers can be got out of von Neumann ordinals (or Zermelo ordinals, or . . .). To say that $2 < 3$ is to say that if x is equinumerous with the $\{\emptyset, \{\emptyset\}\}$ and y is equinumerous with $\{\emptyset, \{\emptyset\}, \{\emptyset, \{\emptyset\}\}\}$, then x is equinumerous with a proper subset of y. To say that $2 + 2 = 4$ is to say that if two disjoint sets x and y are equinumerous with $\{\emptyset, \{\emptyset\}\}$, then their union is equinumerous with $\{\emptyset, \{\emptyset\}, \{\emptyset, \{\emptyset\}\}, \{\emptyset, \{\emptyset\}, \{\emptyset, \{\emptyset\}\}\}\}$. 'Every natural number has a successor' is 'if x is a von Neumann ordinal, the union of x and $\{x\}$ is a von Neumann ordinal'. '2 is prime' says 'if x is equinumerous with $\{\emptyset, \{\emptyset\}\}$, then there are not two sets of cardinality less than 2 but greater than 1 whose cross product is equinumerous with x'. And so on. In practice, these locutions are often simplified even further, but the fact that number theorists have no problem operating inside set theory demonstrates that nothing mathematically important is sacrificed by such translations. The same goes for real numbers and functions.

But it isn't enough simply to do arithmetic and mathematics; our overall theory of the world must also contain a chapter that tells what we are doing and why it works the way it does. This is the descriptive and explanatory theory of our practice required by epistemology naturalized, just the sort of theory, in fact, that we're now trying to construct. Within that theory, explanations of knowledge and reference (or reliability), of puzzles like Wittgenstein's,[46] of the notion of law-like coextensiveness, and so on, might well appeal to objective similarities between individual objects: two triangles are more alike than a triangle and a square; two samples of gold are more similar than either one is to a sample of aluminium; ordered pairs of the form $(x, x + 2)$ are more similar to each other than to a pair of the form $(y, y + 1)$. It remains an open question—for natural science just as for mathematics—whether this distinction between natural and unnatural collections requires a full-blown realism about universals.

[46] See ch. 2, sect. 4, above.

To summarize: for the set theoretic realist, sets have number properties in the same sense that physical objects have length. The further question of the ontological status of these properties is again thoroughly analogous; the mathematical and physical sciences are facing the same metaphysical question. The difference is that the mathematical sciences can offer an answer to part of that question; the set theoretic reductions of arithmetic show that all that is mathematically important about numbers can be said using only sets, while Putnam and others still debate the analogous question in physical science. As for the second question, the metascientific question of what needs to be said in our theory of our respective sciences, the issue is common to both: how are natural kinds, or perhaps objective similarity to be treated? Thus, I take the problem to be a general one, not at all special to the philosophy of mathematics, and that is why I feel justified in leaving it unresolved here.

3. Frege numbers

The previous section on numbers as properties more or less completes what I have to say here on my own view of numbers, but I'm not quite ready to leave the topic entirely. So far, in evaluating the effort to identify numbers with sets, I've concentrated on von Neumann and Zermelo ordinals, giving little consideration to Frege numbers: for example, the Frege number three is the extension of the concept 'equinumerous with the series of natural numbers ending with 2'. I'd like to pause a moment over Frege's idea, not because I intend to modify the theory of numbers already proposed, but because I think attention to Frege numbers will cast some helpful light on what our theory of sets is and is not.

To focus the issue, let me reconsider Benacerraf's argument one last time. We hear of Ernie and Johnny, who learn the von Neumann and Zermelo ordinals, respectively. Each claims that his sets are the natural numbers. Either version will do: there are no conclusive arguments to decide between them. Assuming there is no third candidate for which there are decisive arguments, the conclusion follows that numbers are not sets.

But is there another candidate? Benacerraf mentions a Frege-style proposal, that the number three is the set of all three-element sets. If numbers are actually properties of sets, as I've suggested, Benacerraf fears this would count as a compelling argument in favour of this set theoretic account over von Neumann's and Zermelo's; thus he is motivated to reject the property view, even on slender grammatical grounds. Only the conviction that numbers are not properties of sets allows him to put the Frege-style option aside, and only then is he satisfied that he need look no further:[47] 'There is little need to examine all the possibilities in detail, once the traditionally favored one of Frege . . . has been seen not to be uniquely suitable' (Benacerraf (1965), 284).

There is a missing premiss in this reasoning, something along the lines that if a property is to be treated set theoretically, then it should be identified with the set of things that exemplify it. There is something to this; the set theorist (usually presupposing von Neumann's numbers) speaks not of 'evenness', but of the set of even numbers, not of 'primehood', but of the set of primes. In terms of this methodology, Frege-style sets are the natural choice.

But, though I've embraced the property theory, I've advocated a different line on how these properties ought to be treated within set theory, namely, in terms of comparisons with a standard measuring rod like the von Neumann ordinals. While this approach allows us to say everything we want to say, it leaves us without anything in set theory to properly call 'the numbers'.[48] I think this is as it should be. I've argued that numbers are really properties, in the sense that the theory of numbers, arithmetic, is a theory of number properties. If 'numbers' are 'that which forms the subject matter of arithmetic', then numbers are properties. But if 'numbers' are 'the referents of number words', then there are no numbers. 'One', 'two', 'three', and so on, may enjoy the superficial grammar of names, but they are really just another measuring rod like the Zermelo ordinals.[49] We should resist the urge to find referents for

[47] Notice that Benacerraf doesn't rule out a candidate for identification as the numbers simply because it has superfluous properties; the Frege-style sets have those. Rather he rules out candidates (with superfluous properties) for which there are no good arguments.

[48] Granted the usual conventions, the set theorist will call $\{\emptyset, \{\emptyset\}\}$ 'two', but on on my view, this is like calling the standard metre in Paris 'the metre'.

[49] Compare Benacerraf (1965), 292: 'Questions of identification of the referents of number words should be dismissed as misguided in just the way that a question about the referents of the parts of a ruler would be seen as misguided.' Despite Benacerraf's rejection of the property view, there are many points of contact between the position advocated here and that sketched in the final pages of his paper. I won't try to sort out our agreements and disagreements.

these words, and the temptation to treat number properties as sets is just one version of this misguided impulse.

Still, Frege numbers are of interest, even if it is wrong to say that they are the natural numbers.[50] The extension of 'equinumerous with . . .' is a collection; set theory is our theory of collections; it's natural to ask how this particular collection fits in. Asking this question will tell us something about the nature of sets.

One of the strongest arguments against the claim that Benacerraf's Frege-style sets are a suitable set theoretic version of the Frege numbers is that they don't exist.[51] We've seen that this is true because, for example, new three-element sets are formed at every stage of the iterative construction, so there is no stage at which the set of all three-element sets is formed. This collection is too big to be a set; in standard terminology, it is a 'proper class'. Other collections, like the collection of all sets, are also proper classes. Thus the common wisdom is that the set of all three-membered sets would be a good candidate for the role of a Frege number were it not for the unfortunate fact that it is too big to exist. But I think this underestimates the drawbacks of the set of all three-membered sets.

To see this, consider: ordinary arithmetic is used to count physical objects, perhaps sets of these (like the two pairs of shoes), perhaps occasionally even sets of sets of these, and so on, but never sets of more than finite rank. This is the content of my insistence that arithmetic is part of the theory of hereditarily finite sets. Eventually, of course, we go beyond ordinary arithmetic; we try to number things like the sets of equinumerous sets of real numbers, and then we are doing higher set theory.[52] Just for the moment, I'd like to suggest that this expansion of our arithmetical theory is non-trivial, that—just as the move from ordinary distances to astronomical ones requires a change in our notion of space from Euclidean to non-Euclidean—when we demand that our numbers count more complicated, infinitary things, we are asking for more complicated numbers.[53] On this picture, there *are* Frege-style sets

[50] Readers of Maddy (1981) and (1983) will recognize a change of heart here. I have Michael Resnik to thank for the insight that the very complexity of the theory of the 1983 article suggests it is not a theory of ordinary numbers.

[51] See Benacerraf (1965), 284.

[52] Cantor's continuum hypothesis, to be discussed in the next chapter, says that there are exactly two sets of equinumerous infinite sets of reals.

[53] Notice, these new numbers are not more complicated in that they are infinite—I'm still talking about finite numbers—they are just more complicated in that the finite sets they number can have infinite sets in their transitive closures.

that correspond to the Frege numbers of ordinary arithmetic. The trouble arises only when we try to expand our arithmetic to count objects of higher set theory, but perhaps the trouble is due to the vagaries of higher set theory and not to the simpler numbers of ordinary arithmetic, with which we began.

Now suppose this separation of infinitary versus non-infinitary arithmetic can be maintained. If the only problem with Frege-style sets as surrogates for Frege numbers were their size, we could then identify the 'small Frege numbers', the Frege numbers of ordinary arithmetic, with the corresponding Frege-style sets; that is, the small Frege number three would be the set of all three-membered hereditarily finite sets. What I want to claim is that even here the Frege-style sets are unsuitable.

The small Frege numbers count physical objects (and sets of these, etc.), and presumably the physical objects in our world fluctuate: new animals are born; old stars explode; trees are converted into tables and chairs. Correlated with these shifts in the population of physical objects are shifts in the population of sets: when the cub is born, it is a member of sets that didn't exist before; when the old star explodes, sets also vanish; sets of trees give way to new sets of tables and chairs. Finally, these fluctuations in the population of sets bring about fluctuations in the population of three-membered, hereditarily finite sets, so the set of all these, our Frege-style candidate for the small Frege number three, is now this set, now that. This is not to say that the Frege-style set changes its membership from time to time—sets are fully determined by their members, so they can't do that—rather, the proposal that the small Frege number three be identified with the Frege-style set of all three-element, hereditarily finite sets is the proposal that it be identified with one set after another, with different sets at different times.[54]

This is not satisfactory. If we really thought Frege numbers were the numbers, we might well insist that whatever three is, it is the same thing today as it was yesterday; if we thought the number words had referents, they should have the same referents today as tomorrow. But even if we reject (as I have) the theses that numbers are objects and that number words have referents, I think we should still object to this account of the Frege numbers themselves. When we asked after the collection of all three-membered sets, we didn't want now this collection, now that, depending on the

[54] This argument is an adaptation of Hambourger's (1977), with temporal considerations replacing modal ones.

vagaries of physical existence. What we had in mind was a collection whose membership is allowed to change from time to time while it stays the same, that is, the collection which collects, at any given time, the three-element sets in existence at that time. Such a collection, even if it is small enough, is not a set.

Frege numbers, then, are collections that are not sets. Large Frege numbers are too large to be sets, so they are proper classes. Small Frege numbers, though small enough to be sets, are individuated differently, so they are classes, too, without being proper. Those inclined to insist on referents for the number words will need a theory of sets and classes. Even those immune to this temptation will find the distinction between these two types of collections important to the clarity of their theory of sets. And there remains the possibility that classes might do important work in set theory itself.[55] Let me conclude this section, then, with a brief look at the distinction between sets and classes.

At the beginning of this chapter, I sketched two foundational worries that contributed to the development of the modern theory of sets. The historical path from the troubles with the calculus, through Cantor and Dedekind's work on the real numbers, to Zermelo's axiomatization and beyond, has been a mainly mathematical development whose vicissitudes eventually led to a well-articulated picture of collections formed sequentially, from previously given elements, in a hierarchy of stages: the iterative conception of set. At each stage, new collections are formed with complete freedom, without concern for any method of construction. Finite combinatorics tell us that there is a unique subcollection of a finite collection for every way of saying yes or no to each individual element. Carrying this notion into the infinite, subcollections are 'combinatorially' determined, one for every possible way of selecting elements, regardless of whether there is a specifiable rule for these selections. This is the mathematical notion of a collection: a collection formed combinatorially, in a series of stages that make up the iterative hierarchy.[56]

[55] For example, in non-demonstrative arguments for new set theoretic hypotheses. See Maddy (1988*a*) for some examples.

[56] This picture appears in Zermelo (1930). The combinatorial idea is explicit in Bernays (1935). As indicated above, Enderton (1977), 7–9, Boolos (1971), and Shoenfield (1977) give modern presentations. Of course, the temporal and constructive imagery is only metaphorical; sets are understood as objective entities, existing in their own right.

The second historical trail is the one that begins with Frege's logicism and continues through Russell and Whitehead's effort to salvage some aspects of that idea.[57] Here the original goal was to found arithmetic, and the collections involved are extensions of things more or less like predicates.[58] This is a very different conception: the entire universe is simply divided into two piles, depending on whether or not each thing satisfies the given predicate. This contrasts twice with the iterative picture: we are free to collect absolutely any things, regardless of whether they are all available at some stage, but we are not free to collect combinatorially, without recourse to a rule of any kind. The Cantorian, mathematical, collection is a set; the Fregean, logical, collection is a class.[59]

Collections of these two types differ in several other ways, one of which we've already noted: classes can be larger than sets. Thus, for example, there is a class of all sets—the extension of the predicate 'is a set'—but it is not a set, because there is no stage at which it can be formed. Such a class is a proper class, because it is too large to be coextensive with any set. On the other hand, because classes can only be formed when there are suitable predicates, there may well be more sets of real numbers than there are classes of reals. This depends, of course, on the details of our theory of predicates, but it is hard to imagine how the existence of a predicate for every set at each stage could be guaranteed without somehow presupposing the combinatorial notion of set theory.[60]

We've also seen that a class can change its membership, while a set cannot. The collection of things with a certain property can vary in membership from time to time, but as long as it is identified as the extension of the appropriate predicate, it remains the same class. A set, on the other hand, is completely identified by its membership, so the collection of things with a certain property is now one set, now another.

[57] In fact, Russell and Whitehead's position (1913) is a hybrid between the two notions considered here: extensions, which depend on predicates, are formed in stages. A similar amalgam is proposed by Keith Devlin in support of the axiom candidate V = L (see ch. 4, sect. 4, below). Gödel (1944), 464, traces the ideas behind this axiom to Russell.

[58] Here I'm including Frege's concepts, Russell's propositional functions, my scientific properties, etc. I won't distinguish between these in what follows.

[59] I discuss this contrast in more detail, with more historical considerations, in Maddy (1983). See also Parsons (1974*a*; 1977), and Martin, 'Sets versus classes' (unpublished).

[60] Notice that this counts against efforts, remarked on earlier, to mimic set theory in property theory.

A novel and striking difference is one that shows up in the structure of the membership relation. A class can be a member of another class—the class of von Neumann ordinals is a member of the class of infinite collections—just as sets are members of one another. But the class of von Neumann ordinals isn't the only member of the class of infinite collections—there are infinitely many infinite collections—so the class of infinite collections has the distinction of being self-membered. Of course, no set can be self-membered, because all its members must be formed at stages before that at which it is formed.

And finally, classes, unlike sets, lead to paradox. We've seen that some collections are self-membered—e.g. the class of all infinite collections—while others are not—e.g. any set. 'Being non-self-membered' seems a perfectly unobjectionable predicate, satisfied by some collections, not satisfied by others, so there ought to be a class that is its extension. But this is Russell's paradox in much the same form as he first presented it to Frege:[61] is the class of non-self-membered collections self-membered or not? And again, on the iterative conception, all sets are non-self-membered, so the Russellian set, the set of all non-self-membered sets, is the set of all sets. But there is no such set, because there is no stage at which it could be formed, and thus, no paradox.

Most contemporary set theory is done without explicit mention of classes. Theories that do attempt to encompass both sorts of collections generally separate sets from proper classes on the basis of size, without taking into account the fundamental differences between sets and small classes. Sometimes classes are not allowed as members of any further collections, which avoids paradox, but seems restrictive and artificial;[62] other times they are allowed membership in further collections, but these collections are formed in stages, disallowing self-membership, and are determined com-

[61] Again, see his letter to Frege, Russell (1902).

[62] One system like this is von Neumann–Bernays–Gödel (see von Neumann (1925), Bernays (1937), Gödel (1940)), in which there is a class for every first-order formula with quantifiers ranging only over sets. This allows the familiar Zermelo–Fraenkel axioms to be condensed into a finite list, but has little effect beyond this metamathematical simplification. Morse–Kelley (see Kelley (1955) and Morse (1965)) is a stronger system, allowing quantification over classes in formulas that determine classes, but it still disallows classes as members and has the added difficulty that its classes look like little more than another layer of sets that was somehow left out of the hierarchy. See Drake (1974), 16–17, Fraenkel, Bar-Hillel, and Levy (1973), ch. 2, § 7.

binatorially, which raises serious questions about how these additional layers of non-set collections really differ from sets.[63]

A parallel situation arises in the theory of truth,[64] with sentences like 'Everything I've ever said is false.' If everything else I've ever said *is* false, then this statement is paradoxical. Russell gives us a predicate that can't have an extension; here we have a sentence that can't have a truth value. Parallel to the class theoretic solution that disallows classes as members, we could insist that the question of truth not be raised for sentences involving the notion of truth; this escapes paradox, but again it is too restrictive and artificial.

Another solution, parallel to regimenting classes into stages, requires that the notion of truth be typed: truth_1 applies only to statements that don't involve the notion of truth at all; truth_2 only applies to statements involving truth_1, and so on. Thus 'false' in 'Everything I've ever said is false' is not-true_n for some n, and 'everything' only ranges over statements involving truth_{n-1}. In both cases—class theory and truth theory—the diagnosis is that we can't survey some entire category—all classes, all statements—but only one level or type at a time, and in both cases, the restrictions square poorly with the pre-theoretic notions.

Thus the paradoxes of truth and class theory lead to unpalatable hierarchies in which statements can't refer to themselves and classes can't be self-membered. An alternative available in both cases is to allow gaps, that is, statements that are neither true nor false, classes of which some items are neither members nor non-members. Kripke showed how such a system would go for truth theory,[65] and I have proposed an analogous theory of classes.[66]

To a certain extent, this approach produces the desired results: the class of infinite collections is self-membered; the class of non-self-membered collections is neither self-membered nor non-self-membered. Frege numbers, large and small, can be defined in various ways, and the small ones at least are fairly well behaved. Still, the context of three-valued logic—true, false, and neither—

[63] For example, Ackermann (1956) seems to add several layers of combinatorially determined 'classes'. (See Fraenkel, Bar-Hillel, and Levy (1973), ch. 2, §7.7.) Reinhardt (1974), in a descendant of Ackermann's system, reinforces this picture by explicitly assuming the axiom of foundation for classes. (Foundation disallows self-membership, among other class-like pathologies. See Enderton (1977), 206.)

[64] For further discussion of the parallel, see Parsons (1974*b*).

[65] See Kripke (1975).

[66] In Maddy (1983).

makes the system awkward, so I think the jury must still be considered out on the question of a workable theory of classes in general and of Frege numbers in particular.[67]

[67] Maddy (1984*c*) contains some modifications and developments of the system, as well as answers to one of the open questions of Maddy (1983): a fixed point theorem of William Tait implies that the construction does not reach a fixed point. Robert Flagg has since demonstrated what Tait conjectured, namely, that if the construction is carried out over a standard model of ZF, then the fixed point will be the first admissible ordinal greater than the least upper bound of the ordinals of the ground model. Maddy (1984*c*) also contains some sobering information on the prospects for axiomatization. See Feferman (1984*a*) for a compendium of related systems.

4
AXIOMS

1. Reals and sets of reals

The epistemology of compromise Platonism follows Gödel's in being two-tiered: the most primitive truths are intuitively given, obvious; the more theoretical hypotheses are justified extrinsically, by their consequences, by their ability to systematize and explain lower-level theory, and so on. In Chapter 2, I sketched the set theoretic realist's version of intuition, a neurologically based phenomenon that produces firmly held elementary beliefs and provides them some preliminary level of justification. The remainder of our evidence for the principles we choose as axioms must come from theoretical sources; the time has come to look into the structure of this second type of mathematical justification and to confront the questions it raises.

Because extrinsic arguments involve more advanced levels of mathematical theorizing, they bring us face to face with more esoteric set theoretic matters than have heretofore been relevant. As this is obviously unavoidable, I beg the indulgence of my non-mathematical reader. Indeed, I hope she might come away with a greater appreciation for why Hilbert and others refuse to be budged from Cantor's paradise.

The story begins with the mathematical concerns that led Cantor there in the first place.[1] I'll turn to axiomatics in the next section.

Cantor's dissertation was in number theory, but when he arrived at his first university teaching job in 1869, one of his senior colleagues posed him a problem in analysis:[2] consider functions from reals to reals that can be represented by infinite trigonometric series; are

[1] In ch. 3, sect. 1, above, I put greater emphasis on Dedekind's and Frege's foundational motivations. Cantor's, as we'll see, were more strictly mathematical.

[2] In mathematics, 'analysis' means the study of real and complex functions, which includes the calculus and its foundations and extensions. In this discussion of Cantor's career and its antecedents, I follow Dauben (1979).

such representations unique? Eduard Heine, Cantor's colleague, had proved uniqueness under certain special circumstances, but a general solution eluded him, just as it had his fellow analysts Dirichlet, Lipschitz, and Riemann. Within months, Cantor had obtained the desired result: if a function $f(x)$ is given by a trigonometric series that converges for every value of x, then that representation is unique.

But Cantor didn't stop there. Typical of related work in the theory of functions was the generalization of such theorems by allowing a number of 'exceptional points', for example points at which the series is not required to converge. Heine's partial results had allowed finitely many exceptions, but Cantor, inspired by the work of Hankel, hoped to accommodate infinitely many. The method depended crucially on the distribution of the exceptional points. Everyone knew that infinitely many points contained in a bounded interval will accumulate around at least one point.[3] Cantor could see how to deal with infinitely many exceptional points if they had exactly one accumulation point; with a little effort, he could see how to allow for finitely many accumulation points; indeed, his method would still work on an infinite set of accumulation points as long as that set had only one accumulation point of its own, or for that matter, finitely many accumulation points of its own; and so on. The outlines of the generalization he had in mind were there, but Cantor needed a way to formulate his most lenient condition on exceptional points with precision.

He quickly realized there was no hope of defining such a complicated set of points without an accurate theory of the real numbers themselves; it was this problem that led him to his account of reals in terms of fundamental sequences. He was then able to formulate the requirement on exceptional points appropriate for the generalization of his uniqueness theorem. But what an odd set of points it was: infinite, and quite complex, yet still somehow small enough, or well-behaved enough, in relationship to all the reals, to do no damage! This apparently got Cantor to wondering how continuous sets like the reals relate to seemingly smaller, discrete infinite sets like the natural numbers.

Meanwhile, their shared interest in the theory of real numbers had brought Dedekind and Cantor into correspondence. In a letter

[3] An accumulation point of a set of points has points of that set arbitrarily close to it. for example, 1 is an accumulation point of $\{1/2, 2/3, 3/4, 4/5, \ldots\}$.

to his friend, Cantor raised a nagging question: can the real numbers be brought into one-to-one correspondence with the naturals? Admittedly, there seem to be more reals than naturals, but then, there seem to be more rationals that naturals, too, and Cantor had shown that there were not.[4] He wrote:[5]

> At first glance one might say no, it is not possible, for [the set of natural numbers] consists of discrete parts while [the set of real numbers] builds a continuum; but nothing is won by this objection, and as much as I am inclined to the opinion that [the set of naturals] and [the set of reals] permit no such unique correspondence, I cannot find the reason, and while I attach great importance to it, the reason may be a very simple one.

Dedekind had no easy answer. Cantor replied:

> I raised the question because I have considered it for a number of years and have always found myself doubting whether the difficulty it gave me was subjective or whether it was due to the subject itself. Since you write that you are also in no position to answer it, I may assume the latter.

Shortly thereafter, Cantor found his theorem: the correspondence is impossible; there are, in this sense, more real numbers than there are naturals.[6]

If there are so very many points on the line, how many might there be in a plane, in a three-, or four-, or *n*-dimensional space? This next question to Dedekind took longer to answer, perhaps because Cantor so firmly expected spaces of higher dimension to have more points. When he finally produced a one-to-one correspondence between the line and the plane, he was moved to remark: 'I see it, but I don't believe it!'[7] But there it was. Every infinite set he had considered so far had either the cardinality of the reals or the cardinality of the naturals.

This discovery served to refocus attention right where it had started: on sets of reals. Cantor wrote:[8]

> And now that we have proved, for a very rich and extensive field of [sets], the property of being capable of correspondence with the points of a

[4] See Enderton (1977), 130.

[5] This quotation and the next are drawn from Dauben (1979), 49–50.

[6] See Enderton (1977), 132. A set is 'countable' if it is no larger than the naturals; otherwise it is 'uncountable'.

[7] Dauben (1979), 55.

[8] This and the next quotation from Cantor (1878) are translated by Jourdain in the introduction to his edition of Cantor (1895/7), 45.

continuous straight line . . . the question arises . . . Into how many and what classes (if we say that [sets] of the same or different [size] are grouped in the same or different *classes* respectively) do [infinite sets of reals] fall?

Cantor had an opinion:

By a process of induction, into the further description of which we will not enter here, we are led to the theorem that the number of classes is two . . .

This conjecture, that every infinite set of reals is either countable or of the cardinality of the continuum, has come to be called Cantor's 'continuum hypothesis' (CH).

Cantor's best effort in the direction of a proof of this conjecture involved the notion of a perfect set. A closed set of reals is one that contains all its accumulation points; a perfect set is closed, and every one of its points is an accumulation point.[9] Perfection played a key role in Cantor's analysis of continuity itself, and eventually he was able to show that every non-empty perfect set is equinumerous with the continuum. After a false start pointed out by Ivar Bendixson, Cantor proved that every closed set of reals can be decomposed into a countable set and a (possibly empty) perfect set, which implies that the continuum hypothesis is true for closed sets. (From now on, I'll assume that perfect sets are non-empty by stipulation.) At this point, Cantor was optimistic about the possibility of generalizing this result, called 'the Cantor–Bendixson theorem':[10] 'In future paragraphs it will be proven that this remarkable theorem has a further validity even for [sets of reals] which are not closed . . .'. Of course, a complete generalization would constitute a proof of Cantor's continuum hypothesis in its entirety.

Characterizing sets of exceptions or singularities wasn't the only problem that brought analysts of this period up against questions about sets of real numbers. From the eighteenth century on, a wide range of considerations—from vibrating strings and heat flow to the foundations of the calculus—consistently pushed mathematicians from narrower to more inclusive notions of function.[11]

[9] For example, the set consisting of the points between 0 and 1 inclusive is perfect. The set consisting of those points plus the single point 2 is closed but not perfect.

[10] Cantor (1883), 244, translated by Dauben (1979), 118. For other discussions of the history of the Cantor–Bendixson result, see Moore (1982), 34–5, and Hallett (1984), 90–2.

[11] This development is traced by Kline (1972), 335–40, 403–6, 505–7, 677–9, 949–54.

What began in Galileo's time with curves and continuous motions developed to Euler's combinations of parts of different curves, then to a series of accounts in terms of ever-widening class of expressions that could legitimately define a function, until mathematicians were finally faced with the idea of a purely arbitrary function as absolutely any correspondence between reals and reals regardless of how it might or might not be expressed by mathematical operations. Odd and seemingly pathological examples proliferated: functions that pass through every value between *a* and *b* without being continuous,[12] continuous functions that aren't differentiable,[13] and even Dirichlet's 'shotgun' function (zero on rationals, one on irrationals) which was nowhere continuous, without either derivative or integral.[14]

Doubts circulated about the soundness of this extremely general notion, and by 1900 there was considerable controversy about the proper extent of the function concept.[15] In an effort to get a responsible handle on the vast range of non-continuous functions, the French analysts René Baire, Emile Borel, and Henri Lebesgue set out to give a systematic classification. Lebesgue's version made use of Borel's earlier hierarchy of sets of reals. The simplest sets of reals are the closed sets, mentioned earlier, and their complements, the open sets.[16] The union of two closed sets is closed, but the union of countably many closed sets may be open[17] or worse.[18] This 'or worse' gives rise to the Borel hierarchy:[19]

Σ^0_1 = the open sets
Π^0_1 = the closed sets
Σ^0_2 = countable unions of closed sets

[12] Due to Darboux. See Kline (1972), 952.

[13] By Riemann, Cellérier, and Weierstrass between 1854 and 1875. See Kline (1972), 955–6.

[14] In 1829. See Kline (1972), 950.

[15] See Monna (1972).

[16] Equivalently, a set is open if it contains on open interval ($\{x \mid a < x < b\}$) around each of its points.

[17] For example, the union of the closed intervals $[1/n, (n-1)/n]$ is the open interval $(0, 1)$. (By notational convention, square brackets indicate that the endpoints are included, round brackets that they are not.)

[18] For example, the union of the closed intervals $[0, (n-1)/n]$ is neither closed nor open.

[19] The Borel sets were introduced by Borel (1898) as sets resulting from the closed sets using complement and countable intersection. Lebesgue (1905) introduced a hierarchy, though not this one, which is due to Hausdorff (1919). It continues to generate new sets of reals until the first uncountable ordinal. (For limit ordinals λ, Σ^0_λ = countable unions of sets in the Π^0_αs for a $\alpha < \lambda$.) For the basic theory of Borel sets, see Kuratowski (1966), §§5, 6, and 30, or Moschovakis (1980), ch. 1.

$$\Pi_2^0 = \text{complements of } \Sigma_2^0 \text{ sets}$$

.

.

.

$$\Sigma_{\alpha+1}^0 = \text{countable unions of } \Pi_\alpha^0 \text{ sets}$$
$$\Pi_{\alpha+1}^0 = \text{complements of } \Sigma_{\alpha+1}^0 \text{ sets}$$
$$\Delta_\alpha^0 = \text{sets that are both } \Sigma_\alpha^0 \text{ and } \Pi_\alpha^0$$
$$\text{Borel sets} = \text{the union of the } \Sigma_\alpha^0\text{s}$$

Lebesgue's hierarchy of 'Borel functions' was defined in terms of these simple sets of reals, and he proved it equivalent to another hierarchy given earlier by Baire.[20]

The Borel sets, despite their complexity, turn out to be fairly well behaved. Consider, for example, the perfect subset property: a collection of sets of reals is said to have the perfect subset property if every infinite set in the collection is either countable or contains a perfect subset.[21] Thus the Cantor–Bendixson theorem says that the closed sets of reals have the perfect subset property. Paul Alexandroff (1916) made good on Cantor's hunch that this result could be generalized by extending it to include all Borel sets. Another property of interest to analysts is separability: two sets A and B are separated by a set C if A is a subset of C and B is disjoint from C. Wacław Sierpiński (1924) proved that disjoint Borel sets in Π_α^0 can be separated by a Borel set in Δ_α^0, and gave applications of this fact to the general theory of functions.

Lebesgue was also instrumental in the isolation of a second important collection of sets of reals. The impetus this time came from the theory of integration, another field filled with perplexities at the time. Among the numerous pathological functions under scrutiny were seemingly unobjectionable examples that turned out not to be integrable using the state-of-the-art Riemann integral. To extend the concept of integration, Lebesgue needed a new gauge of the size of a set of reals, not a generalization of number, like Cantor's cardinality, but a generalization of length that could take that concept from intervals to more complex sets. For this purpose,

[20] To see the connection, notice that a function f is continuous if it has the following property: whenever A is an open set, $f^{-1}[A] = \{x \mid f(x) \in A\}$ is also open. Lebesgue defined a function g to be Σ_α^0 iff $g^{-1}[A]$ is Σ_α^0 whenever A is Σ_α^0. Thus the Σ_1^0 functions are the continuous ones. Baire's version of this hierarchy appears in Baire (1899).

[21] Recall I'm assuming perfect sets are non-empty.

he developed the notion of Lebesgue measure,[22] and even Dirichlet's function became integrable. Since closed sets are Lebesgue measurable, and the collection of Lebesgue measurable sets is closed under complement and countable union, it follows that Borel sets have another nice property: they are all Lebesgue measurable.

The next step in this development, the last one I'll touch on here, was also precipitated by Lebesgue, but this time by an uncharacteristic error. One of his analyses of Baire functions included a 'trivial' lemma that the projection of a Borel set is Borel.[23] This isn't true, but the slip went unobserved for a decade, until Mikhail Suslin, a young student in Moscow, burst into his professor's office with the news. Together, student and professor, Nikolai Luzin, established the elementary properties of these 'analytic' sets,[24] which led, some years later, to the introduction of a new hierarchy of sets of reals:[25]

$$\begin{aligned}
\Sigma^1_0 &= \text{the open sets}\\
\Pi^1_0 &= \text{the complements of } \Sigma^1_0 \text{ sets}\\
\Sigma^1_1 &= \text{the projections of } \Pi^1_0 \text{ sets}\\
\Pi^1_1 &= \text{the complements of } \Sigma^1_1 \text{ sets}\\
&\ \ \vdots\\
\Sigma^1_{n+1} &= \text{the projections of } \Pi^1_n \text{ sets}\\
\Pi^1_{n+1} &= \text{the complements of } \Sigma^1_{n+1} \text{ sets}\\
&\ \ \vdots\\
\Delta^1_n &= \text{sets that are both } \Sigma^1_n \text{ and } \Pi^1_n\\
\text{projective sets} &= \text{the union of the } \Sigma^1_n
\end{aligned}$$

The relationship between Borel and projective sets was cinched by Suslin, who showed that the Borel sets are the Δ^1_1 sets.

Despite their added complexity, projective sets inherit some of

[22] In Lebesgue (1902). For an elementary exposition, see Williamson (1962).

[23] Instead of sets of points on the line, think of sets of points in the plane; Borelness and so on can be defined for these sets just as easily. Then the projection of a Borel set in the plane is, so to speak, the shadow it casts on the x-axis.

[24] See Suslin (1917) and Luzin (1917).

[25] Introduced by Luzin (1925) and Sierpiński (1925). For the classical theory of projective sets, see Kuratowski (1966), §§ 38–9, or Moschovakis (1980), §§ 1E, ch. 2, and parts of ch. 3.

the regularity properties of Borel sets. Lebesgue measurability and the perfect subset property for analytic or Σ^1_1 sets were immediately established by Luzin and Suslin and separability came a few years later.[26] Separability was extended to Π^1_2 by P. Novikov in his (1935), but after that, the best efforts stalled: separability remained open past Π^1_2, measurability beyond $\Sigma^1_1$and Π^1_1, and for the perfect subset property, as Luzin remarked:[27] 'There remains here only one important gap: we do not know if every uncountable complement of an analytic set [i.e. every uncountable Π^1_1 set] has the [cardinality] of the continuum.' There the matter stood for some years.

Thus, for all the thoroughly admirable foundational goals pursued by Frege and Dedekind, the deepest contributions to the modern mathematical theory of sets, those of Georg Cantor, were inspired almost exclusively by mathematical concerns, particularly concerns arising from analysis. I have tried in this section to give a hint of how Cantorian set theory grew out of the study of real functions and to sketch in the sorts of questions that arose naturally in that context. Let me turn now to how and why this naïve mathematical theory came to be axiomatized.

2. Axiomatization

To understand the impulse that led to the axiomatization of set theory, we must return to Cantor and his continuum problem. Around the time he was making his optimistic prediction, quoted earlier, about generalizing the Cantor–Bendixson theorem, another letter to Dedekind contains the first hint of what was to become a corner-stone of his theory of infinite numbers, namely, the concept of well-ordering.[28]

Again the catalyst was his earlier work on sets of singular points. In order to describe the process of taking accumulation points of sets of accumulation points of sets of accumulation points, etc., he had proceeded as follows from a given point set A:

[26] In Luzin (1927).
[27] Luzin (1925), as translated by Hallett (1984), 108.
[28] See Moore (1982), 40–1. The letter in question is dated 5 Nov. 1882.

$$A_0 = A$$
$$A_1 = \{x \mid x \text{ is an accumulation point of } A_0\}$$
$$\vdots$$
$$A_{n+1} = \{x \mid x \text{ is an accumulation point of } A_n\}$$
$$\vdots$$
$$A_\omega = \text{the intersection of the } A_n\text{s}$$
$$A_{\omega+1} = \{x \mid x \text{ is an accumulation point of } A_\omega\}$$

and so on. Now, in the early 1880s, his interest shifted from the derived sets themselves to the subscripts. Why shouldn't there be $\omega + \omega$ after all the $\omega + n$s, and $(\omega + \omega) + 1$ after that? Here is a sequence that carries into the transfinite, and after any batch of entries, there is always a next.[29]

Considered in terms of cardinality, this transfinite sequence of ordinals yields a wonderful bonus. The collection of all countable ordinals is not itself countable. In fact, as Cantor had shown by the late 1890s,[30] the cardinality of this set is the very next infinite cardinality after that of the natural numbers. Thus, he called the latter $\aleph_0$ and the former $\aleph_1$. And the set of ordinals of cardinality $\aleph_1$ has cardinality $\aleph_2$, and so on, so the infinite sequence of ordinals yields an infinite sequence of cardinal numbers as well. Cantor extended the arithmetic operations—plus, times, exponentiation—from the finite numbers into the infinite, and showed that the cardinality of the continuum is $2^{\aleph_0}$. By this point, then, the continuum hypothesis had become: $2^{\aleph_0} = \aleph_1$. But, for all Cantor's efforts, it remained unproved.

Then, in 1904, at the Third International Congress in Heidelberg, came a shock; Julius König read a paper that purportedly showed the continuum hypothesis to be false. In particular, König argued that the continuum could not be well-ordered at all, and *ipso facto*, that it could not be put in one-to-one correspondence

[29] Technically, a well-ordering of a set A is an ordering in which every non-empty subset of A has a least member. This produces the idea in the text: as we run through the elements of A, at any point there is a least member of the elements we haven't listed yet. See Enderton (1977), 172–3.

[30] See his last major work, Cantor (1895/7).

with the well-ordering $\aleph_1$. A contemporary report describes the scene this way:[31]

> For Cantor to claim that every set can be well-ordered and, in particular, that the continuum has the second [infinite] cardinality was a kind of dogma that was part and parcel of what he knew and believed in set theory. Consequently König's address, which culminated in the proposition that the continuum could not be an aleph (hence could not be well-ordered either), had a stunning effect, especially since its presentation was extremely elaborate and precise.

Apparently Cantor was less upset by the proof itself, of which he was sceptical, than he was by what he saw as his public humiliation before his colleagues.[32] His scepticism, at least, was well taken; one of König's assumptions, a 'theorem' of Felix Bernstein, turned out to be incorrect. This was pointed out to the members of the Congress, on the day directly after the presentation of König's proof, by Ernst Zermelo.

But a doubt remained. As early as 1883, just after the above-described letter to Dedekind in which the notion is introduced, Cantor wrote that:[33]

> The concept of a *well-ordered set* turns out to be essential to the entire theory of point-sets. It is always possible to bring any *well-defined* set into the *form* of a *well-ordered* set . . . this law of thought appears to me to be fundamental, rich in consequences, and particularly marvelous for its general validity . . .

By 1895, he realized the need for a proof of this fundamental principle, and set out to find one. His letter to Dedekind of 1899 contains one attempt.[34] Now, though König's attack had been unsuccessful, it brought home the possibility that not only the continuum hypothesis, but the well-ordering principle itself might one day be overthrown.

[31] Schoenflies, as quoted in van Heijenoort (ed.) (1967), 192.

[32] See Dauben (1979), 247–50, 283, for discussion of this episode.

[33] Cantor (1883), as translated by Moore (1982), 42. For an analysis of Cantor's thinking on well-ordering, see Hallett (1984), § 3.5.

[34] In fact, Cantor's argument is very close to the one usually given today (see e.g. Drake (1974), 56); it goes through with minor modifications once Zermelo's axiom has been identified. Apparently it was the (unnecessary) use of proper classes—his 'inconsistent multiplicities'—that disturbed Cantor. In any case, neither he nor others (including Hilbert) who saw the argument at the time were convinced by it. In 1903, when it was rediscovered by Jourdain, Cantor refused have his version published. See Moore (1982), §§ 1.6 and 2.1 for details.

Before 1904 was over, a proof was presented, not by Cantor, but again by Zermelo.[35] The surprisingly short argument depends on a novel assumption:

> . . . that even for an infinite totality of sets there are always mappings that associate with every set one of its elements . . .

By way of defending this principle of choice—the function 'chooses' one element from each set—Zermelo admits that

> This logical principle cannot, to be sure, be reduced to a still simpler one . . .

but in its favour:

> it is applied without hesitation everywhere in mathematical deduction . . .

As an illustration, he cites the seemingly obvious fact that a set can't be divided into more non-empty disjoint parts than it has members, a fact whose proof also depends on choice.

In this unassuming manner, Zermelo proposes that mathematicians accept a principle that is simple, that was so obvious to previous theorists that they used it unconsciously, and that is necessary to prove various important and natural results (like the well-ordering principle, the theorem that all infinite cardinalities are alephs, and the partition principle just noted). He could hardly have been prepared for the storm of controversy that ensued. Gregory Moore's extraordinary history of this period traces the complex reaction through its independent manifestations in France, Germany, Hungary, England, Italy, Holland, and the United States.[36] Choice and a welter of other set theoretic principles, including well-ordering and unlimited comprehension, were suddenly up for grabs, embraced here, denied there. It was in defence of his principle of choice and his proof of well-ordering that Zermelo was driven to axiomatize the practice of set theory.

Zermelo's defence of the principle of choice and his axiomatization of set theory appear in two papers written within days of one

[35] Zermelo (1904). This short paper began as a letter to Hilbert one month after the Congress and was published later that year. All the quotations in this paragraph come from p. 141.

[36] Moore (1982), ch. 2. My brief account of the controversy over choice draws heavily on Moore's work.

another in the summer of 1907.[37] The first of these contains a spirited response to critics of what was now his axiom of choice. He admits again, as he had in 1904, that the principle has not been proved, but points out: 'even in mathematics *unprovability* . . . is in no way equivalent to *nonvalidity*, since, after all, not everything can be proved, but every proof in turn presupposes unproved principles' (Zermelo (1908*a*), 187). Thus, even his opponents must rely on unproved axioms. How, then, are these axioms justified?—'by pointing out that the principles are intuitively evident and necessary for science . . .' (Zermelo (1908*a*), 187). From our set theoretic realist's perspective on mathematical evidence, Zermelo is recognizing both intrinsic supports—in terms of 'intuitive evidence'—and extrinsic supports—in terms of the role of the axiom in overall scientific theorizing. He proposes to apply these criteria to choice.

To confirm the intuitiveness of his axiom, Zermelo cites historical evidence:

> That this axiom, even though it was never formulated in textbook style, has frequently been used, and successfully at that, in the most diverse fields of mathematics . . . is an indisputable fact . . . Such an extensive use of a principle can be explained only by its *self-evidence* . . . (Zermelo (1908*a*), 187)

Cantor, Dedekind, and the other early set theorists had passed over numerous uses of choice in various forms without comment, usually without noticing it themselves.[38] Unconscious applications of the principle can also be found in analysis, particularly in the work of Baire, Borel, and Lebesgue touched on in the previous section.[39] This surely constitutes some evidence for its obviousness, and hence, for its intuitiveness, though the initial protest against Zermelo's linguistic formulation remains to be explained (see p. 123 below). Zermelo anticipates the objection that evidence for intuitiveness should not be counted as evidence for truth:

> No matter if this self-evidence is to a certain degree subjective—it is surely a necessary source of mathematical principles . . . and Peano's [40] assertion that it has nothing to do with mathematics fails to do justice to manifest facts. (Zermelo (1908*a*), 187)

[37] These are Zermelo (1908*a*) and (1908*b*).

[38] For examples, see Moore (1982), § 1.4.

[39] See Moore (1982), §§ 1.7 and 4.1. For the role of the axiom in the work of Suslin and Luzin, see §§ 3.6 and 4.1.

[40] Peano was among the critics of Zermelo's axiom. See Moore (1982), § 2.8.

Here he appeals to the undeniable fact that intuitiveness is often taken, in practice, as evidence for truth. This isn't enough (as observed in the third section of Chapter 2), but the additional considerations cited there can be brought to bear.

Turning to extrinsic supports, Zermelo gives his version of the role of set theory in our overall theory:

> Set theory is that branch of mathematics whose task is to investigate mathematically the fundamental notions 'number', 'order', and 'function', taking them in their pristine, simple form, and to develop thereby the logical foundations of all arithmetic and analysis; thus it constitutes an indispensable component of the science of mathematics. (Zermelo (1908*b*), 200)

Set theory is essential to mathematics, especially arithmetic and analysis. If we add to this the Quine/Putnam-style claim that mathematics is essential to our theory of the world, we have an indispensability argument for set theory. The question then becomes, what version of set theory is essential to mathematics, and in particular, does that version include choice? Zermelo argues that it does:

> the question that can be objectively decided, whether the principle is *necessary for science* [by which Zermelo means the science of mathematics], I should now like to submit to judgement by presenting a number of elementary and fundamental theorems and problems that, in my opinion, could not be dealt with at all without the principle of choice. (Zermelo (1908*a*), 187–8)

He goes on to list a series of theorems from set theory and analysis, from which he concludes:

> Now so long as the relatively simple problems mentioned here remain inaccessible to Peano's [choiceless] expedients, and so long as, on the other hand, the principle of choice cannot be definitely refuted, no one has the right to prevent the representatives of productive science from continuing to use this 'hypothesis'—as one may call it for all I care—and developing its consequences to the greatest extent, especially since any possible contradiction inherent in a given point of view can be discovered only in that way . . . principles must be judged from the point of view of science, and not science from the point of view of principles fixed once and for all. (Zermelo (1908*a*), 189)

If choice produces a better, more effective theory than choiceless mathematics, Zermelo counsels that we opt for choice and jettison any unscientific prejudice that stands in our way.

Zermelo's contention that mathematics without choice would be 'an artificially mutilated science'[41] was substantially confirmed in the years following. The implicit uses of choice in analysis were uncovered gradually, and it is now clear that the theory of real functions and Borel and projective sets sketched in the previous section would change significantly without a weak version of choice called 'dependent choice' and would collapse completely without the even weaker 'countable choice'.[42] The theory of transfinite cardinal numbers, on the other hand, could hardly survive without the full axiom of choice.[43] And, to cite just one more example, the importance of the principle in algebra is dramatically demonstrated in the history of Bartel van der Waerden's classic textbook. The first edition, published in 1930, included the axiom and its already considerable store of algebraic consequences.[44] The book itself stimulated further productive research along these lines, but Dutch opponents convinced van der Waerden to omit choice in his second edition and return to more familiar methods. Algebraists were appalled by this 'mutilated' version of their discipline, and the axiom and its consequences were reinstated by popular demand in the third edition in 1950.[45]

In the case of the axiom of choice, then, we have our first example of an extrinsic defence of a set theoretic hypothesis, beginning with a straightforward indispensability argument: our best theory of the world requires arithmetic and analysis, and our best theory of arithmetic and analysis requires set theory with at least the axiom of dependent choice. Beyond this pure Quine/Putnamism, the compromise Platonist finds the sort of intra-mathematical arguments that Gödel anticipates.[46] First, as

[41] Zermelo (1908*a*), 189.

[42] The principle of countable choice says that there is a choice function for any countable collection of non-empty sets; dependent choice says that these choices can be made in such a way that each depends on the previous one. See Moore (1982), 103 and 325, and Moschovakis (1980), 423 and 445, for discussions of the role of these principles.

[43] For example, the principle that for any two cardinal numbers κ and λ, either $\kappa = \lambda$ or $\kappa < \lambda$ or $\lambda < \kappa$ is equivalent to the full axiom of choice. See Moore (1982), §4.3, and 330–1, for more.

[44] For a discussion of these, see Moore (1982), §3.5.

[45] See Moore (1982), §4.5.

[46] In Gödel (1947/64), 477. All the quotations in this paragraph come from that location.

Sierpiński noted,[47] it has a number of 'verifiable consequences', that is, 'consequences demonstrable without the new axiom, whose proofs with the help of the new axiom, however, are considerably simpler and easier to discover . . .' Second, it yields a 'powerful method' for solving pre-existing open problems, for example the well-ordering question. And finally, it systematizes and greatly simplifies the entire theory of transfinite cardinal numbers, 'shedding so much light upon a whole field . . . that . . . [it] would have to be accepted at least in the same sense as any well-established physical theory'.

These are not, however, the only arguments mathematicians have given in favour of the axiom of choice. To trace the source of the other main line of defence, let me return to the two concepts of collection discussed in the final section of Chapter 3. The mathematical notion, originating in Cantor's thinking about sets of pre-existing points, eventually developed into the full iterative conception of Zermelo (1930). The logical notion, beginning with Frege's extension of a concept, now takes a number of different forms depending on exactly what sort of entity provides the principle of selection, but all these have in common the idea of dividing absolutely everything into two groups according to some sort of rule.

In the early 1900s, these two notions had not yet been distinguished, and this ambiguity is what produced the deepest division over Zermelo's principle. One of the great ironies of this entire historical episode is that the strongest negative reaction to the axiom came from the very group of French analysts—Baire, Borel, and Lebesgue—who unwittingly used it with great frequency and whose work provides part of the basic indispensability argument for at least dependent choice.[48] Yet it is not hard to see how this happened. The efforts of these analysts were originally motivated by their doubts about the extremely general notion of function proposed by Dirichlet and Riemann; instead they concentrated on developing their hierarchies of functions definable by acceptable mathematical means. The conflict here is between the notion of functions as completely arbitrary correspondences, one for each possible combination of pairs of reals, and the notion of functions as transformations, determined by some sort of definition or rule.

[47] In Sierpiński (1918). For discussion, see Moore (1982), § 4.1.
[48] See Moore (1982), §§ 1.7, 2.3, and 4.1.

Transferred into the realm of sets, this is just the contrast between the mathematical and the logical notions of collection.

The true shape of this conflict emerged in the aftermath of Zermelo's first proof, in a series of letters between the three analysts and their opponent, Jacques Hadamard.[49] There Lebesgue writes:

> to define a set M is to name a property P which is possessed by certain elements of a previously defined set N and which characterizes, by definition, the elements of M. . . . The question comes down to this, which is hardly new: *Can one prove the existence of a mathematical object without defining it?* . . . I believe that we can only build solidly *by granting that it is impossible to demonstrate the existence of an object without defining it.* (Baire *et al.* (1905), 314)

Hadamard's position was the opposite:

> . . . Zermelo provides no method to carry out *effectively* the operation which he mentions, and it remains doubtful that anyone will be able to supply such a method in the future. Undoubtedly, it would have been more interesting to resolve the problem in this manner. But the question posed in this way (the effective determination of the desired correspondence) is nonetheless completely distinct from the one that we are examining (does such a correspondence exist?). . . . Can one prove the existence of a mathematical object without defining it? I answer . . . in the affirmative. . . . the *existence* . . . is a fact like any other . . . (Baire *et al.* (1905), 312, 317)

For Lebesgue and the rest, the existence of a choice function, or to put it more simply, a choice set,[50] depends on our having a rule with which to determine what is in the set and what isn't. For Hadamard, what rules we have is irrelevant, purely psychological; a set either exists or it doesn't.

Lebesgue's point of view is obviously most plausible on the logical notion of collection, the notion in fact suggested in the above quotation. In contrast, the mathematical notion, according to which, for any given things, there is a set consisting of any combination of those things, aligns with Hadamard's thinking. Given a collection of non-empty, disjoint sets formed from some batch of things, a choice set will be among the combinations of those original elements. From this point of view, Zermelo's axiom becomes obvious. During the debate over choice, the notion of

[49] Baire *et al.* (1905).

[50] The axiom can be rephrased to say: for any collection of non-empty, disjoint sets, there is a set that contains exactly one element from each of them.

collection in question was still ambiguous, which is why some found the principle obvious and others found it preposterous.

Indeed, the very depth of the conviction on both sides suggests that both notions of collection enjoy some intuitive backing. This would also explain why many opponents of choice continued to use it unawares even after the principle had been isolated and the controversy was joined; recall that intuitions, properly so called, are common to (nearly) all. On this theory, then, Zermelo's historical evidence does support the intuitiveness of the principle, but his linguistic formulation met with protest because the word 'set' therein wasn't always connected up with the appropriate pre-linguistic intuitive beliefs.

Fleshing this idea out requires a few more speculative wrinkles to the perceptual/intuitive story told in Chapter 2, but let me indicate how it might go. We learn to perceive 'something with a number property'. This concept develops in the course of extensive childhood experience with manipulating and rearranging medium-sized physical objects, so it is inextricably linked to finite combinatorial notions, for example the idea that in any order, there are still ten pennies, or that absolutely any proper subcollection of the pennies will number fewer than ten. This underlies the mathematical, combinatorial notion. On the other hand, on most occasions of counting, the child counts things collected under some umbrella: the boys in the garden, the pennies in my pocket. Even seemingly random collections are for the most part spatially circumscribed: the things in this box. This aspect of numerical experience underlies the logical notion. Thus the intuitive concept of 'that which has a number' contains elements of both notions.

Only later do we realize that these two notions can come apart. In finite cases, it might be argued that any combinatorial possibility is also determined by a property—the property 'being this thing or that thing or . . . ' through a finite list of the members—though this approach will be problematic, for example on accounts of scientific properties as 'natural' collections. But infinite cases raise more pressing doubts. Is there such a rule for determining membership in each and every combinatorially determined subset of the natural numbers? It isn't obvious that there is. Yet the combinatorial ideal suggests that each and every such subset can be counted, just as finite collections can be counted even when no non-trivial membership rule is available. This clash of intuitions—between 'every collection is collected by some property' and 'any collection, with a common property or not, can be counted'—leads to confusion.

Eventually, theoretical considerations take over. Sets rather than classes make the most workable and fruitful mathematical entities, and functions understood as arbitrary mappings provide an important flexibility that narrower notions of function cannot equal. By these means, we are led to the conclusion that being collected under one umbrella was an accidental feature of the 'things with number properties' of our childhood experience, and not an essential feature of all 'things with number properties'. Not surprisingly, then, it is to such theoretical facts that Hadamard ultimately appeals:

> From the invention of the infinitesimal calculus to the present, it seems to me, the essential progress in mathematics has resulted from successively annexing notions which, for the Greeks or the Renaissance geometers or the predecessors of Riemann, were 'outside mathematics' because it was impossible to describe them. (Baire *et al.* (1905), 318)

In the years since this controversy raged, the mathematical notion has been developed and accepted, because of its effectiveness, and the Hadamard position has prevailed. Thus D. A. Martin, one of our leading contemporary set theorists, writes:

> much of the traditional concern about the axiom of choice is probably based on a confusion between sets and definable properties. In many cases it appears unlikely that one can *define* a choice function for a particular collection of sets. But this is entirely unrelated to the question of whether a choice function *exists*. Once this kind of confusion is avoided, the axiom of choice appears as one of the least problematic of the set theoretic axioms. (Martin, 'Sets versus classes' (unpublished), 1–2)

This last pro-choice argument, then, is that objections to choice are based on the wrong notion of collection. It depends on both the intuitive evidence for choice assuming the mathematical notion and the theoretical evidence that the mathematical notion is the correct one.

I have concentrated on the axiom of choice because its fascinating history provides the clearest illustration of the interplay between intuitive and theoretical supports for set theoretic hypotheses, but a similar analysis can be given for each of the currently accepted axioms of Zermelo–Fraenkel set theory. I won't do this here,[51] because my goal is illustrative rather than exhaustive, but let me at least indicate the extreme variety of that list. Pairing—for any two things there is a set with exactly those members—and Union—any sets can be combined into a set with all their members as members—have been cited earlier[52] as examples of nearly unadorned intuitions. In contrast, the axiom of infinity—there is an infinite set—proclaims the bold and revolutionary hypothesis that led Cantor into his paradise and the rest of us with him. There is nothing obvious about it, but it launched modern mathematics, and the success and fruitfulness of that endeavour provides its purely theoretical justification.

3. Open problems

By the mid-1930s, the fundamental assumptions underlying set theoretic practice had been codified into a simple axiomatic system, ZFC, which was strong enough to imply the known theorems of classical number theory and analysis and pre-axiomatic set theory. The well-orderability of the reals was provable within this theory, leaving the continuum hypothesis in Cantor's preferred form: $\aleph_1 = 2^{\aleph_0}$. But the question remained open, as did those raised by Suslin and Luzin: is there an uncountable Π^1_1 set with no perfect subset? Are all Σ^1_2 and Π^1_2 sets Lebesgue measurable? Can disjoint Σ^1_3 or Π^1_3 sets be separated?

Developments in logic during the twenties and early thirties, especially Gödel's completeness and incompleteness theorems of 1930 and 1931,[53] raised the possibility of a new sort of proof—a proof of unprovability—and it was in this direction that progress on these open problems was first made. The most dramatic result of Gödel's work in the late thirties was his demonstration that the

[51] Maddy (1988*a*) contains a summary of many of these arguments.

[52] In Ch. 2, sect. 3, above.

[53] Gödel (1930; 1931). The completeness theorem establishes that every logically valid formula is provable. (See Enderton (1972), ch. 2, for a now-standard proof using the alternative methods of Henkin (1949).) The first incompleteness theorem shows that there are sentences expressible in the language of ZFC that aren't provable or disprovable from ZFC. The second incompleteness theorem gives an example of such a sentence, namely, the one expressing the consistency of ZFC. (See Enderton (1972), ch. 3.)

continuum hypothesis can not be disproved from the axioms of ZFC.[54] At the same time, Gödel noted some consequences for analysis which were finally proved some time later by John Addison:[55] in ZFC, it cannot be proved that all uncountable Π^1_1 sets have perfect subsets or that all Δ^1_2 sets are Lebesgue measurable. On the question of separation, Addison used similar methods to show that ZFC does not imply either the separability of Σ^1_3 sets or the non-separability of Π^1_3 sets.[56]

The possibility remained that the axioms of ZFC would be enough to establish the continuum hypothesis as true, but Gödel for one did not expect this. This conjecture was based partly on evidence that ZFC is too weak to decide the question at all, and partly on his strong conviction that the continuum hypothesis is in fact false: 'certain facts (not known at Cantor's time) . . . seem to indicate that Cantor's conjecture will turn out to be wrong . . .' (1947/64), p. 479). These facts consist of 'highly implausible consequences of the continuum hypothesis', of which Gödel lists several. Thus we find Gödel arguing against Cantor's conjecture on extrinsic grounds, in terms of its purportedly undesirable consequences.

The question is, what makes these particular consequences undesirable? Many of them depend on the idea that sets of reals which are large in number should not also be small in measure-theoretic terms. There might well be some intuitive belief lurking behind this vague principle, but if so, it is extremely undependable: an elementary theorem of measure theory shows that there are uncountable sets of Lebesgue measure zero; the standard example goes back to Cantor. Given that the suggested principle has been discredited, it might be argued that Gödel is relying on some other, more dependable intuition. If so, we should expect most other set theorists to share his views, but they don't. Martin, for example,

[54] Unless ZFC is inconsistent; anything can be both proved and disproved in an inconsistent system. (Terminology: to disprove is to prove the negation, so a proof that CH cannot be disproved is a proof that its negation cannot be proved.) See Gödel (1940).

[55] In Addison (1959). Other partial results by Mostowski and Kuratowski were destroyed during the war. See Addison's paper for the complicated history of these theorems. Moschovakis (1980), § 5A, presents Addison's results.

[56] In Addison (1958).

writes:[57] 'While Gödel's intuitions should never be taken lightly, it is . . . hard for some of us to see why the examples Gödel cites are implausible at all' (Martin (1976), 87). The final, most likely possibility is that Gödel is relying, not on intuition, but on his mathematical experience, exercising the sort of theoretical judgement that produces the natural scientist's hunch that a theory of this sort rather than that is the kind that ought to work. Unfortunately, Gödel's efforts to pin down his ideas have since proved unsuccessful.[58]

But whatever his reasons, Gödel was correct in his prediction that the continuum hypothesis would be shown not to follow from ZFC. This result was finally established by Paul Cohen in 1963.[59] Solovay then extended Cohen's method to treat questions from analysis: ZFC cannot disprove that all uncountable Π^1_1 sets have perfect subsets or that all Σ^1_2 and Π^1_2 sets are Lebesgue measurable.[60] This inconclusive picture was completed by Leo Harrington,[61] who used Cohen's method to show that ZFC does not yield the separation property for Σ^1_3 or Π^1_3. Thus these various problems are open in an entirely new sense of the word; provably, they cannot be decided on the basis of the standard assumptions of set theory.

Many have been provoked to philosophical extremes by the thought of questions this open. The if-thenist, for example, simply declares all such problems to be solved; what we wanted to know, after all, was whether or not the continuum hypothesis and the rest follow from ZFC. But whatever machinations might be available to her opponents, the Platonist's position is clear. In Gödel's words:

It is to be noted, however, that on the basis of the point of view here

[57] See also Martin and Solovay (1970), 177. Moore (forthcoming) attributes a similar lack of agreement to Cohen. An exception to this rule is Nyikos (forthcoming), who agrees with Gödel that at least one of these examples is extremely implausible.

[58] For an account of these efforts, see Moore (forthcoming), § 6.

[59] The published version is Cohen (1966).

[60] See Solovay (1970). Solovay's result actually depends on the relatively weak additional assumption that the existence of an inaccessible cardinal (see the next section) cannot be refuted in ZFC. Moschovakis writes, 'In the present context this is surely a reasonable assumption' (1980), p. 284), and I know of no dissent from this view.

[61] See Moschovakis (1980), 284.

adopted, a proof of the undecidability of Cantor's conjecture from the accepted axioms of set theory . . . would by no means solve the problem. For if the meanings of the primitive terms of set theory [based on the iterative conception] are accepted as sound, it follows that the set-theoretical concepts and theorems describe some well-determined reality, in which Cantor's conjecture must be either true or false. Hence its undecidability from the axioms being assumed today can only mean that these axioms do not contain a complete description of that reality. (Gödel (1947/64), 476)

For the set theoretic realist, the world consists of physical objects, sets of these, sets of physical objects and sets, and so on, through the transfinite levels of the iterative hierarchy. There is a fact of the matter about the cardinality of the set of Dedekind cuts—a set of pairs of sets of sets of pairs of von Neumann ordinals—and it is the set theorist's job to discover it.[62]

And how might this elusive fact be ascertained? Gödel's answer, and the set theoretic realist's, is that we need to find new axioms, axioms we can justify just as Zermelo's axioms were justified, by a combination of intrinsic and extrinsic considerations.[63] This approach is seconded by our current set theorists. For example, Martin writes: 'Although the ZFC axioms are insufficient to settle CH, there is nothing sacred about these axioms, and one might hope to find further axioms . . .' (Martin (1976), 84). Considerable work has been done on this project in recent years, some of which will be described in the next section.

In another attack on Gödel's ideas, Chihara takes this situation as evidence against Platonism.[64] Gödel's and Cohen's results show that both ZFC + CH and ZFC + not-CH are consistent theories, perhaps equally worthy of investigation.[65] Chihara concludes that

For Gödel, . . . the proliferation of set theories poses the thorny problem of

[62] Of course, set theory is as fallible as any other science, and it could turn out that the continuum question is based on faulty presuppositions, but there is no conclusive reason to believe this now.

[63] See Gödel (1947/64), 476–7.

[64] See Chihara (1973), 63–5. All quotations in this paragraph and the next come from this location.

[65] In fact, neither of these theories is much studied, as such, because neither CH nor not-CH is considered a viable axiom candidate. Neither is intuitive and neither is sufficiently fruitful to merit acceptance on extrinsic grounds, though CH is better off than not-CH in this regard: Sierpiński (1934) derives eighty-two propositions from CH, none of which is known to be settled by not-CH (see Martin and Solovay (1970), 143).

determining which of the many set theories [one for each possible cardinality of the continuum] is the one that most truly describes the real world of sets. (Chihara (1973), 65)

Now no Platonist would deny that the continuum hypothesis poses a 'thorny problem'—it has engaged many of the best set theoretic minds since Cantor's—but it isn't immediately clear why this casts doubt on Platonism. After all, scientific realism leaves us with many thorny problems, from the shape of the universe to the existence of a free quark, and no one counts this as evidence that there is no objective physical world.

On the Platonist's view, there is a real and extremely difficult problem about the cardinality of the continuum. Chihara seems to hold that this fact counts in favour of alternative philosophies of set theory for which the continuum problem presents no such challenge. For example, he suggests as a 'reasonable option' his 'mythological Platonism', which takes the continuum hypothesis to be analogous to 'Hamlet's nose is three inches long', that is, to be neither true nor false. But how reasonable would it be for the physicist to solve the question of the free quark by adopting a philosophy of physics according to which it is no longer a problem? In fact, thorny problems are the life-blood of science, its motivator, and set theory is no different from the rest.

If we are to look towards new axioms for a solution to the continuum problem, it is worth asking in which direction that solution might be expected. Opinion on this matter is divided, but opinion there is. No one pretends to have anything resembling conclusive evidence for any alternative, but a brief look at the range of considerations offered will give the flavour of the debate.

Cantor, of course, held the continuum hypothesis to be true, and occasionally, during his many efforts, even believed that he had proved it.[66] One of the forces behind his strong conviction may have been his confidence that the partial solution contained in the Cantor–Bendixson theorem could be generalized.[67] To a certain extent, we've seen that this confidence was well placed: Cantor–Bendixson was extended, most dramatically by Suslin to the Σ^1_1 sets. Thus, despite the fact that there are far fewer Σ^1_1 sets than there

[66] For examples, see Moore (1982), 42–5, and Hallett (1984), 92.

[67] Cantor had other reasons, too. See Maddy (1988*a*), 490–2, and the references cited there.

are arbitrary sets of reals,[68] this partial result affirming the continuum hypothesis for a wide range of sample sets might be taken as confirmation for the hypothesis in general.

Unfortunately, whatever confirming evidence Suslin's result may have promised, it is severely undercut by the details of the argument itself. What he actually shows is that every uncountable Σ^1_1 set has a perfect subset. The trouble is that some uncountable sets don't have perfect subsets, so there is actually no hope of generalizing Suslin's result on Σ^1_1 sets to all sets of reals. In Martin's words: 'Thus, while our simple [Σ^1_1] sets have the cardinalities required by CH, this is so because they have an *atypical* property, the perfect subset property' (Martin (1976), 88). In Cantor's defence, it should be noted that most of the sets he was familiar with were Σ^1_1 at worst, and Bernstein's theorem on uncountable sets without perfect subsets didn't appear until 1908.

If Cantor's reasons for believing the continuum hypothesis are ultimately unpersuasive, Gödel's reasons for disbelieving it have also drawn few converts. By contrast, the sentiments of another major player in this drama proceed along lines that many seem to find more plausible. Cohen's thinking depends on a contrast between two ways in which larger cardinals can be generated from smaller ones.

One method builds up from below. Cantor's original procedure for building ever larger ordinals depended on three principles of generation.[69] The first allows the passage from one number to its successor, from 2 to 3, from 3 to 4. There is no largest number in this series, so Cantor's second principle generates their 'limit', the next number after them all, that is, ω. From here, the first principle yields $\omega + 1$, $\omega + 2$, etc., and the second, $\omega + \omega$. And so on. But all ordinals generated by these processes are still countable. The third principle tells us that after all ordinals of a certain cardinality, there is a next, in this case, ω_1. This ordinal is uncountable; its cardinal number is $\aleph_1$.

These methods produce a sequence of infinite cardinal numbers—$\aleph_0$, $\aleph_1$, $\aleph_2$, $\aleph_3$, etc.—but they cannot take us beyond this point. To build up further from below, we need the axiom of replacement: given any set, if each of its elements is replaced by something else, the result is still a set. Thus if $\{0, 1, 2, 3, \ldots\}$ is a

[68] There are $2^{\aleph_0}$ Σ^1_1 sets of reals and $2^{2^{\aleph_0}}$ sets of reals altogether.

[69] Cantor (1883). See Dauben (1979), 96–9, or Hallett (1984), §2.1, for discussion.

set, as the axiom of infinity guarantees, and 0 is replaced by the set of finite ordinals, 1 by the set of countable ordinals, 2 by the set of ordinals of size $\aleph_1$, and so on, we have the set whose elements have the cardinalities $\aleph_0$, $\aleph_1$, $\aleph_2$, and so on. If we take the union of all these sets, as the union axiom says we can, the result is a set whose cardinality is the next largest after $\aleph_0$, $\aleph_1$, $\aleph_2$, etc. This is $\aleph_\omega$. Obviously, this process can be continued.

The second way of generating larger cardinalities is very different. Beginning again with the set $\{0, 1, 2, 3, \ldots\}$, this time we take the set of its subsets, as allowed by the power set axiom. One of Cantor's most beautiful theorems shows that the power set of any set has a larger cardinality than the set itself;[70] in this case, that larger cardinality is $2^{\aleph_0}$, also the cardinal of the set of reals.[71] Taking the power set of this power set yields a set of cardinality $2^{2^{\aleph_0}}$, and so on.

Cohen's idea is simply that power set is stronger than any principle for building up from below.[72] In his book establishing the unprovability of the continuum hypothesis, he writes:

> A point of view which the author feels may eventually come to be accepted is that CH is *obviously* false. . . . $\aleph_1$ is the set of countable ordinals and this is merely a special and the simplest way of generating a higher cardinal. The set [of subsets of the natural numbers] is, in contrast, generated by a totally new and more powerful principle, namely the Power Set Axiom. It is unreasonable to expect that any description of a larger cardinal which attempts to build up that cardinal from ideas deriving from the Replacement Axiom can even reach [a set of size $2^{\aleph_0}$]. Thus [$2^{\aleph_0}$] is greater than $\aleph_n$, $\aleph_\omega$, $\aleph_\alpha$, where $\alpha = \aleph_\omega$, etc. This point of view regards [the power set of the set of natural numbers] as an incredibly rich set given to us by one bold new axiom, which can never be approached by any piecemeal process of construction. (Cohen (1966), 151)

This line of thought harmonizes with various others insisting that the continuum hypothesis places an artificial and unwarranted restriction on the number of reals.[73] Most such thinkers—they include Martin[74]—feel that the continuum is likely to be quite large compared with $\aleph_1$. This sets them apart from Gödel, who, while

[70] Cantor (1891). See Enderton (1977), 132–3.

[71] See Enderton (1977), 149.

[72] Dauben (1979), 269, traces this way of thinking to Baire.

[73] See Maddy (1988*a*), § II.3.4. This paper gives a more complete and detailed list of the various arguments for and against Cantor's hypothesis.

[74] See e.g. his 'Projective sets and cardinal numbers' (unpublished), 2. Some of Martin's views are reported in Maddy (1988*a*), § V.4.

rejecting Cantor's hypothesis, seemed to lean towards a relatively small continuum of size $\aleph_2$.[75]

So the axiomatization of set theory produced a range of problems more open than had previously been possible, that is, problems neither provable nor disprovable from the accepted assumptions in the field. The most famous of these is Cantor's continuum hypothesis, but various others appeared among the natural questions asked by analysts in the twenties and thirties. From the Platonist's perspective, there is good reason to believe that these questions nevertheless have unambiguous answers; some even proffer opinions about what those answers might be. The hope is for new, strongly supported axioms that will resolve these difficult issues.

4. Competing theories

In the current set theoretic landscape, two opposing theoretical approaches dominate efforts to solve the profoundly open problems described above. My goal in this section is to give an overview of each, with special attention to perceived strengths and weaknesses. In the final section of this chapter, I'll turn to the philosophical questions raised by this controversy.

The most complete and concise of these two theories stems from Gödel's proof that the continuum hypothesis cannot be disproved from the axioms of ZFC. To show this, Gödel described a simplified world of sets in which all the axioms, and hence all consequences of the axioms, are true. But in this world, $\aleph_1 = 2^{\aleph_0}$, so the negation of the continuum hypothesis cannot be proved from ZFC.[76]

This simplified world of Gödel's is arranged in a hierarchy of stages, just like the standard iterative hierarchy, and there is one stage for every ordinal number, again in imitation of the standard picture. The difference is that at any given stage, instead of forming all possible subsets of what has been given so far, in Gödel's world

[75] See Moore (forthcoming), esp. § 6.

[76] See Drake (1974), ch. 5, or Jech (1978), §§ 12–13, for textbook presentations of this argument.

one adds only those subsets explicitly definable by predicative[77] formulas. This procedure yields the constructible universe, called L, and the claim that the constructible universe is the real universe, written V = L, is the axiom of constructibility. Gödel's work shows that adding the axiom of constructibility to ZFC can't introduce any contradictions that weren't already present in ZFC alone.

So one live theoretical option is just that: add V = L to the assumptions of ZFC. This move produces a theory so powerful that the axiom of choice is no longer needed; it can be proved. In fact, not only is every set well-orderable—a condition equivalent to choice[78]—but the entire universe can be arranged in a giant well-ordering—a condition called 'global choice'. This holds in L because the relevant formulas can be well-ordered at each stage, which produces in turn a well-ordering of the sets introduced at that stage. The well-orderings of the stages are then strung together to produce a well-ordering of the entire constructible universe. The axiom also has striking effects in the area of cardinal arithmetic; it implies not only the continuum hypothesis, but the generalized continuum hypothesis as well, that is, that $\aleph_{\alpha+1} = 2^{\aleph_\alpha}$, for all ordinals α.

The open questions from analysis are also decided in ZFC + V = L. Addison's unprovability results, like Gödel's, are established by showing that the negations of the propositions in question follow from the axiom of constructibility. In the constructible universe, there is an uncountable Π^1_1 set with no perfect subset, a non-Lebesgue-measurable Δ^1_2 set, and separation holds for Π^1_3, Π^1_4, Π^1_5, and so on. All these follow from the existence of a particularly simple well-ordering of the reals; as a subset of the plane, it is Δ^1_2.

So V = L answers all our outstanding questions. Indeed, further elaborations, largely due to Ronald Jensen, settle nearly all important set theoretic questions, and some from other branches of mathematics as well.[79] Here is an axiom that clearly provides

[77] That is, formulas that only refer to sets formed at previous stages.

[78] See Enderton (1977), 151–4, 196–7, 199.

[79] Devlin (1977) makes the case for V = L's efficacy both inside and outside set theory. The importance of Jensen's contribution to constructibility theory comes out in the historical notes to Devlin's compendium (1984) and in the introduction: 'without his work there would have been practically nothing to write about!' (p. viii).

'powerful methods for solving problems'.[80] It is also a safe assumption; as remarked above, it engenders no contradictions that wouldn't already follow from ZFC alone. There are even those who find it a 'natural' assumption, beginning with Gödel, who introduced it this way:

> The proposition [V = L] added as a new axiom seems to give a natural completion of the axioms of set theory, in so far as it determines the vague notion of an arbitrary infinite set in a definite way. (Gödel (1938), 557)

Keith Devlin goes even further, claiming constructibility to be 'closely bound up with what we mean by "set"', but he defends this by identifying sets with extensions of properties rather than combinatorially defined collections.[81]

Despite the remark just quoted, Gödel soon came to reject V = L, and despite its strengths, constructibility today has more detractors than supporters. The most fundamental reason for resistance to the axiom is implicit in my crude sketch: instead of forming all possible subsets of what has been given so far, one adds only those subsets explicitly definable by predicative formulas. This requirement on subsets clearly violates the combinatorial idea that every possible collection be formed, regardless of whether there is a rule for determining which previously given items are members and which are not. Devlin, to bolster his case, concocts a clever compromise between the logical and the mathematical notions of collection—he proposes that extensions of properties be formed in stages,[82] thus avoiding inconsistency—but the combinatorial idea he rejects stands at the end of a clear historical trend in mathematics, from functions and collections determined by rules towards functions and collections determined arbitrarily. Most set theorists have adopted the full iterative conception and thus find the axiom of constructibility an artificial restriction. For example, Moschovakis writes:

> The key argument against accepting V = L . . . is that the axiom of constructibility appears to restrict unduly the notion of *arbitrary* set . . . there is no

[80] Gödel (1947/64), 477.

[81] See ch. 3, sect. 3, above for this distinction. The quotation is from Devlin (1977), p. iv. The relevant notion of set is sketched on pp. 13–18. Fraenkel, Bar-Hillel, and Levy (1973), 108–9, also cite a range of considerations in favour of V = L.

[82] See Devlin (1977), 27–8.

a priori reason why every subset ... should be definable ... (Moschovakis (1980), 610)

And Gödel: '[V = L] states a minimum property. Note that only a maximum property would seem to harmonize with the concept of set ...' (Godel (1947/64), 479). Many others express opinions along these lines.[83]

Further arguments against the axiom of constructibility focus on its consequences. Of course, anything that might count as a reason for disbelieving the continuum hypothesis would likewise count against V = L; it may be that Gödel's change of mind was partially motivated in this way. Though anti-constructibility arguments of this form are sometimes offered, more common and concrete objections involve the consequences of V = L for the occurrence of so-called pathologies low down in the projective hierarchy, among the fairly simple sets of reals. For example, the axiom of choice implies that the real numbers can be well-ordered, but there is every reason to suppose that the simplest such well-ordering (as a subset of the plane) is an extremely complex set. Provably, it cannot be as simple as Σ^1_1, but in the presence of the axiom of constructibility, it is Δ^1_2. This means that, beginning from a set too simple to be a well-ordering of the reals, such a well-ordering can be obtained by one application of complementation and one application of projection, a prospect which seems highly unlikely to many. Indeed many set theorists feel that such a choice-generated oddity should not appear anywhere among the projective sets. The same sort of thinking applies to the uncountable Π^1_1 set with no perfect subset and the Δ^1_2 non-Lebesgue-measurable set; though choice guarantees that such sets exist, they should not be so simple.[84]

The second live theory begins from a style of axiom for which Gödel had high hopes, namely the large cardinal axiom. The first of these, the axiom of inaccessible cardinals,

roughly speaking, means nothing else but that the totality of sets obtainable by use of the procedures of formation of sets expressed in the other axioms forms again a set (and, therefore, a new basis for further applications of these procedures) ... (Gödel (1947/64), 476)

[83] See e.g. Drake (1974), 131, Scott (1977), p. xii, and Wang (1974*a*), 547.

[84] Opinions of this sort can be found in Martin (1977), 811, (1976), 88, and 'Projective sets and cardinal members', p. 2; Moschovakis (1980), 276; and Wang (1974*a*), 547.

The idea is that the two main operations for generating new sets from old postulated by ZFC—replacement and power set—are not enough to exhaust all the ordinals.[85] Obviously this thinking is ripe for generalization. Such axioms, according to Gödel, show that ZFC 'can be supplemented without arbitrariness by new axioms which only unfold the content of the [iterative conception of set] . . .' (Gödel (1947/64), 477). He proposed large cardinal axioms as a cure for the sort of open problems we've been considering:

> It is not impossible that . . . some completeness theorem would hold which would say that every proposition expressible in set theory is decidable from the present axioms plus some true assertion about the largeness of the universe of all sets. (Gödel (1946), 85)

By the sixties, large cardinals of ever-increasing size were a boom industry.

The most striking application of inaccessibles themselves is Solovay's theorem, mentioned above, applying Cohen's method to questions in analysis. This important work presupposes the irrefutability, if not the existence, of an inaccessible cardinal. Other small large cardinals have consequences for Borel sets,[86] but the most dramatically effective large cardinal axiom postulates the existence of a much larger measurable cardinal.[87] To give just a modest hint of its size, there are inaccessibly many inaccessible cardinals smaller than the first measurable.

The strongest arguments for the assumption of measurable cardinals are extrinsic ones, most notably Dana Scott's discovery that it implies $V \neq L$.[88] Thus the strong considerations against the axiom of constructibility all return as extrinsic supports for the existence of a measurable cardinal. And, once again, Solovay coaxed out results in analysis. He showed that in addition to refuting constructibility, a measurable cardinal yields the preferred results for projective sets: every uncountable Σ^1_2 set has a perfect subset, every Σ^1_2 and Π^1_2 set is Lebesgue measurable, and there is no

[85] Inaccessibles were introduced in Zermelo (1930) and in Sierpiński and Tarski (1930). For the sort of argument in their favour considered here, see also Drake (1974), 267, and Wang (1974*a*), 554. For arguments of other types, see Maddy (1988*a*), 501–4.

[86] See the detailed and innovative work of Harvey Friedman, described in Harrington *et al.* (1985).

[87] Measurable cardinals were first proposed by Ulam (1930). See Drake (1974), chs. 6 and 8, or Jech (1978), ch. 5, for textbook discussions.

[88] Scott (1961).

Δ^1_2 well-ordering of the reals.[89] But for all the good news, there is also bad. The most conspicuous open problem remains so; Solovay and Levy's application of Cohen's method shows that measurable cardinals cannot decide the continuum hypothesis.[90]

Then, in the late sixties, the appearance of a short paper by David Blackwell[91] produced a surge of interest in hypotheses of a completely different sort, hypotheses given in game-theoretic terms. The notion of an infinite game was first introduced by the Polish school in the thirties. Such a game can be based on any set A of real numbers between 0 and 1. Imagine two tireless players who takes turns choosing digits. When they are done (!), they will have constructed an infinite sequence, which can be taken as a decimal expansion, which represents a real number, *r*. If *r* is in A, the first player wins; otherwise the second player wins. The game A is determined if there is a winning strategy for one player or the other. Blackwell used the known fact that all open games[92] are determined to give a new and elegant proof of Luzin's theorem on the separability of Σ^1_1 sets.

This came as a welcome surprise to Moschovakis, Addison, and Martin, all of whom were engaged in efforts to go beyond Novikov's theorem and extend the separation property to higher levels of the projective hierarchy. From the researches of the early analysts, it was known that separation holds for sets at the circled levels:

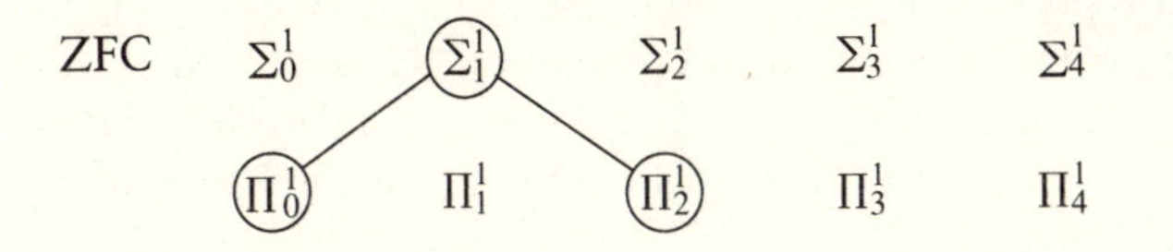

In the constructible universe, Addison had shown that the pattern extends on the Π-side:

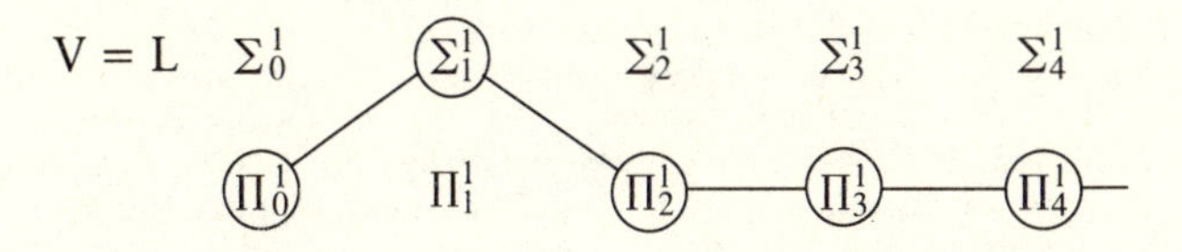

89 Solovay (1969).
90 Levy and Solovay (1967).
91 Blackwell (1967).
92 That is, games whose set of wins for the first player is an open subset of [0, 1].

Even for disbelievers in V = L, this meant that ZFC can't disprove separation for Π^1_3 sets, but for all that was known at the time, it remained possible that ZFC could prove Π^1_3 sets separable. Still, few expected this. A new hypothesis was needed.

In this context, then, Blackwell's proof focused considerable attention on determinacy assumptions. Σ^0_3 sets were known to be determined, a result which Martin later extended to all Borel sets.[93] On the other hand, the axiom of choice implies the existence of a non-determined set.[94] But again, it seems natural to insist that such a choice-generated oddity not appear among the relatively simple sets, for example among the projective sets. Thus projective determinacy—the assumption that all projective sets of reals are determined—was proposed by a wide range of researchers.[95]

Projective determinacy, like the axiom of measurable cardinals, is supported by many of the considerations thought to count against V = L. For example, it extends the results obtainable from a measurable cardinal by guaranteeing not only that every uncountable Σ^1_2 set has a perfect subset, but that every uncountable projective set has a perfect subset; not only that all Σ^1_2 and Π^1_2 sets are Lebesgue measurable, but that all projective sets are Lebesgue measurable; not only that there is no Δ^1_2 well-ordering of the reals, but that there is no projective well-ordering of the reals.[96] Many agree with Martin that these are 'pleasing consequences about the behavior of projective sets'.[97] And the separation question, the problem that inspired this renewed interest in determinacy, was solved by Addison and Moschovakis and independently by Martin;[98] the initial zigzag pattern continues for the length of the projective hierarchy:

[93] The result for Σ^0_3 is in Davis (1964), Martin's in (1975). See Moschovakis (1980), §6F, for a proof of Martin's theorem for the finite levels of the Borel hierarchy. Martin (1985) gives a simplified proof for all Borel sets.

[94] Gale and Stewart (1953). See Moschovakis (1980), 293.

[95] First Solovay and Takeuti, independently, then Addison, Martin, and Moschovakis. (See Addison and Moschovakis (1968), 708–9, Moschovakis (1970), 31, and (1980), 422, 605, and 610–11, Martin (1976), 90, (1977), 814, and 'Projective sets and cardinal numbers', p. 8.) In fact, most of these propose a stronger axiom candidate, quasi-projective determinacy, but I won't go into the exact definition here. See Maddy (1988*a*), 737. Determinacy assumptions were introduced by Mycielski and Steinhaus (1962).

[96] Lebesgue measurability appears in Mycielski and Świerczkowski (1964), and the perfect subset property in Davis (1964). The non-existence of a projective well-ordering follows because a well-ordering is not Lebesgue measurable.

[97] Martin (1976), p. 90.

[98] Addison and Moschovakis (1968) and Martin (1968).

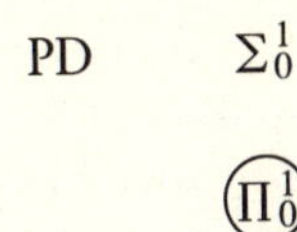

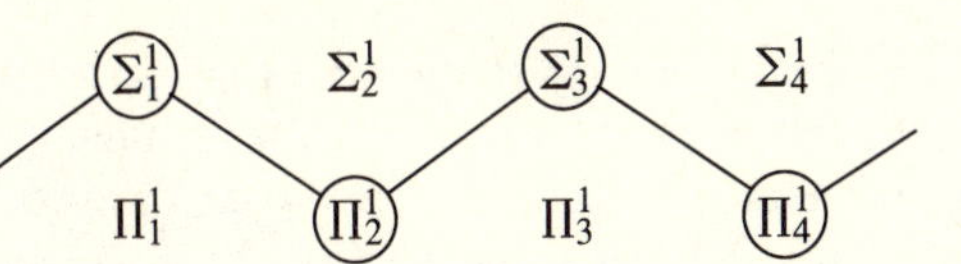

This pattern is considered more natural than the one generated by V = L, if only because /\/\/\/\/\/\/\/\/\/\/\ is a more natural continuation of /\ than /____________.[99]

Other extrinsic evidence for projective determinacy is found in its strong intertheoretic connections with the axiom of measurable cardinals,[100] and in the naturalness of the new game-theoretic proofs:

> One [reason for believing projective determinacy] is the *naturalness* of the proofs from determinacy—in each instance where we prove a property of Π^1_3 (say from [the determinacy of Δ^1_2 sets]), the same argument gives a new proof of the same (known) property for Π^1_1 . . . Thus the new results appear to be natural generalizations of known results and their proofs shed new light on classical [analysis]. (This is not the case with the proofs from V = L which all depend on the [Δ^1_2] well-ordering of [the reals] and shed no light on Π^1_1.) (Moschovakis (1980), 610)

Moschovakis's thick book contains various beautiful and persuasive examples.[101]

But perhaps the most striking feature of determinacy hypotheses, what makes this a particularly fascinating case for the philosopher, is that *all* arguments given in its favour from the mid-sixties until the mid-eighties are extrinsic. Determinacy supporters were quite explicit on this point:

> No one claims direct intuitions . . . either for or against determinacy hypotheses . . . (Moschovakis (1980), 610)

> There is no *a priori* evidence for [projective determinacy] . . . (Martin (1976), 90)

> Is [projective determinacy] true? It is certainly not self-evident. (Martin (1977), 813)

For twenty years, while extrinsic arguments of the sort outlined here developed rapidly, there was no change in the lack of intrinsic support. And yet projective determinacy was still considered a viable axiom candidate.

[99] See Moschovakis (1970), 33–4; Martin (1977), 806, 811, and 'Projectice sets and cardinal numbers', p. 8; Wang (1974*a*), 547, 553–4. For other reasons, see Maddy (1988*a*), § v.1.

[100] See Maddy (1988*a*), § v.2.

[101] See also Maddy (1988), § v.3.

The best hope for something more than purely extrinsic evidence lay in the possibility of deriving determinacy hypotheses from suitable large cardinal axioms.[102] Martin showed, early on, that the determinacy of Σ^1_1 sets is implied by the existence of a measurable cardinal:

> Some set theorists consider large cardinal axioms self-evident, or at least as following from *a priori* principles implied by the concept of set. [The determinacy of Σ^1_1 sets] follows from large cardinal axioms. It is possible that [projective determinacy] itself follows from large cardinal axioms, but this remains unproved. (Martin (1977), 813)

> One way to increase the evidence for [projective determinacy] would be to prove it from large cardinal axioms . . . (Martin, 'Projective sets and cardinal numbers' (unpublished), 8)

Unfortunately, attempts to extend Martin's result made use of cardinals so large that even the most enthusiastic large cardinal theorists were concerned for their consistency.[103]

Before turning to the developments of the mid-eighties, we should pause to ask why deriving determinacy hypotheses from large cardinal assumptions is viewed as providing non-extrinsic support. Obviously no such theorem can provide direct intrinsic support for a determinacy hypothesis; this a hypothesis either has or lacks on its own. What happens in such a case is that the determinacy hypothesis proved inherits the existing supports for the large cardinal assumption from which it is proved. Thus Martin's theorem places the power of arguments for measurable cardinals squarely behind the determinacy of Σ^1_1 sets. But the arguments given above for measurable cardinals were all extrinsic![104] Where is the intrinsic evidence supposed to come from?

The answer is that there are various other arguments for measurable cardinals that don't depend on consequences. In fact, however, I think these arguments rest on ideas that are not happily classified as either intuitive or extrinsic, ideas Martin referred to

[102] Moschovakis (1970), 31, notes Solovay's conjecture that this might be possible, and cites Martin's result as confirming evidence.

[103] Martin's theorem on measurable cardinals and determinacy appears in Martin (1970). Martin (1980) extends the result to Σ^1_2 sets, and Woodin took this line even further. See Maddy (1988*a*), §VI, and Martin and Steel (1989) for discussion of these developments and the very large large cardinals involved.

[104] In fact, its implication of the well-supported Σ^1_1 determinacy is also counted as extrinsic evidence for measurable cardinals.

above as '*a priori* principles implied by the concept of set'. One example has already been given in favour of inaccessibles: the idea that the iterative hierarchy is too large and complex to be exhausted by the simple operations of replacement and power set. The more general idea that motivates all large cardinals is simply that the universe goes on through as many transfinite stages as it can. Various large large cardinals are also defended on grounds arising from the idea that the hierarchy of stages is so complex and rich that it must contain stages that resemble one another in certain ways.[105] For want of a better term, I call these 'rules of thumb'.

The rules of thumb underlying large cardinal axioms are clearly rooted in the iterative conception, which is drawn, I've suggested, from intuition. Why, then, do I avoid attributing intuitive status also to these rules of thumb? The answer is simple: because I think they extend beyond anything that could plausibly be traced to an underlying perceptual, neurological foundation. That sets are formed from previously given things, that they are formed combinatorially, with no concern for rules of formation—these ideas might well have intuitive backing, and there is no doubt they figure centrally in the iterative conception. But if, as I've suggested, the support for the assumption of an infinite stage is purely extrinsic, following from the immense success of modern infinitary mathematics, then part of the standard iterative conception, the part that drives the hierarchy into the infinite and insists that it go on as long as possible, that part is based on the developing methodology of set theory itself, not on simple intuition.

On the other extreme, why shouldn't these principles be counted as extrinsic? Natural science has its own principles of similar generality, for example Maxwell's principle that a law of nature should be valid at all points in space and time.[106] Such principles are indirectly subject to extrinsic support—if they consistently led to ineffective theories, they would eventually be dropped—but this proves nothing; even intuitions must be confirmed by consequences. An unavoidably central aspect of the appeal of these rules of thumb—be they scientific or mathematical—is that they 'seem

[105] See Maddy (1988*a*), § VI.2. Though the arguments for large large cardinals are similar is some structural and philosophical respects to the arguments for small large cardinals, they cannot be said to be equally convincing. Again, Maddy (1988*a*) provides details.

[106] Wilson (1979) discusses this example.

right', even if this seeming is unlikely to enjoy a strictly intuitive basis. If we confine ourselves to an unbiased description of practice, I think we must admit that rules of thumb fall somewhere between the intuitive and the extrinsic.

Leaving aside for now the problem of elucidating and defending the efficacy of this new category of purported evidence, let me pick up the thread of the story of determinacy. Large cardinals are thought to enjoy certain non-extrinsic justifications in terms of various rules of thumb. Thus, if projective determinacy could be proved from such an axiom, its purely extrinsic defence would be enriched by support from this other source. This, then, was the goal.

What happened in the mid-eighties was a fulfilment of this hope. Work of Martin, John Steel, and Hugh Woodin[107] established that projective determinacy[108] can be proved from the existence of a supercompact cardinal. Though supercompacts are much larger than measurables—there are measurably many measurable cardinals below the first supercompact—they can be viewed as a generalization of that notion.[109] Thus the theory ZFC + SC (there is a supercompact cardinal) enjoys all the extrinsic supports of projective determinacy plus any non-extrinsic, rule-of-thumb evidence available for large cardinals in general and for supercompacts in particular.[110] It solves all the outstanding open problems from analysis, and it does so in ways that many find natural. In many respects, then, ZFC + SC presents an attractive alternative to ZFC + V = L.

But what about the continuum hypothesis? If it is true in L, a minimal environment, perhaps it is false in a maximized world of large cardinals. There is some evidence in that direction in results proved from the full (false) axiom of determinacy: if every set is determined, then the reals can be mapped onto $\aleph_2$, $\aleph_{\omega+1}$, $\aleph_{\omega^{(\omega^\omega)}+1}$, and beyond. In the presence of the axiom of choice, this would imply the existence of sets of reals of these cardinalities, badly falsifying the continuum hypothesis, but full determinacy also

[107] See Martin and Steel (1988), Woodin (1988), and Martin and Steel (1989). These papers also contain useful historical information.

[108] Indeed more. The full quasi-projective determinacy, mentioned in n. 95 above, is provable from this large cardinal assumption.

[109] See e.g. Solovay, Reinhardt, and Kanamori (1978), §2. For a textbook discussion of supercompactness, see Jech (1978), 407–13.

[110] For a survey of the latter, see Maddy (1988*a*), §§ VI.1 and VI.2.

implies that choice is false. Indeed, in that strange world of full determinacy, every uncountable set of reals has a perfect subset, so the continuum hypothesis is true in Cantor's original form: all infinite sets of reals have the cardinality of the naturals or the cardinality of the reals themselves. But in Cantor's favoured form—$\aleph_1 = 2^{\aleph_0}$—it remains false; the continuum isn't of size $\aleph_\alpha$ for any α because it can't be well-ordered.

It isn't clear what all this madness means for the real world if there is a supercompact cardinal and the axiom of choice is true. Supporters of ZFC + SC disagree over their expectations for the size of the continuum,[111] not because some support restrictive principle like V = L, but because they disagree over whether $\aleph_1 = 2^{\aleph_0}$ or its opposite is the least restrictive.[112] (Notice, it is possible to read the equation as restricting the number of reals, but also as maximizing the number of countable ordinals.) Efforts are under way to extend the theory to decide the question one way or the other.

I have described two theories, two extensions of ZFC, that cannot both be true. Each theory answers at least the open questions of Luzin and Suslin, and one even decides the size of the continuum. Each enjoys an array of extrinsic supports, supplemented to varying degrees by intuitive and rule-of-thumb evidence, a small portion of which has been described in this summary. The philosophical open question is: on what rational grounds can one choose between these two theories?

5. The challenge

We've seen that set theory arose in response to both foundational and purely mathematical concerns, that it developed as a branch of analysis and as a study in its own right, and that both pursuits produced natural open questions. In the confusion surrounding the paradoxes and Zermelo's controversial proof of the well-ordering theorem, the informal practice was axiomatized, which development had two consequences of concern to us here: the role of

[111] For example, Foreman (1986) proposes 'generic large cardinal' axioms that imply CH. Martin, on the other hand, has conjectures about the relationship between the cardinals in the world of full determinacy and the real world that imply the CH is badly false. See Maddy (1988*a*), § V.4.

[112] See Maddy (1988*a*), §§ II.3.4 and II.3.11.

extrinsic argument in mathematics was crystallized as never before, and it became possible to show that various natural open problems in set theory and analysis were open in a new and stronger sense. The hope of answering these extremely open questions led to the emergence of two competing theories that answer all or most of them and do so in very different ways. The difficult problem facing the set theorists of our day is how to adjudicate between these, that is, how to determine, within the limits of our cognitive capacities, which is more likely to be true.

Proof is obviously the most common source of evidence in mathematics, but even proof must begin from axioms that are not themselves proved. In many circles, the preferred account of our knowledge of axioms is that they are somehow self-evident; in the words of Roderick Chisholm, following Leibniz, 'once you understand it, you see that it is true'.[113] But our brief survey in this chapter shows that even the accepted axioms of ZFC do not enjoy this status, let alone the more controversial axiom candidates like projective determinacy. A new account of our knowledge of axioms and of the evidential role of non-demonstrative mathematical arguments in general is clearly needed.

For the set theoretic realist, non-demonstrative arguments come in two varieties—intuitive and extrinsic—and examination of cases reveals what is probably at least one more—those based on rules of thumb. In Chapter 2, I argued that intuitive support is prima-facie evidence for truth. Perhaps it is not surprising that most of the available intuitive evidence is marshalled in support of the accepted axioms of ZFC. In the dispute between V = L and SC, I've suggested that V = L labours at a modest intuitive disadvantage, because it rejects the combinatorial component of set theoretic intuition. But, given that intuitive evidence is never conclusive, that it needs supplementation by extrinsic supports and can be (indeed has been) overthrown by theoretical counter-evidence, this alone can hardly settle the question in favour of SC.

What's needed is a dependable method for comparing the strength of the non-intuitive, non-demonstrative arguments relevant to this case, some of which were touched on in the previous section. But before we can answer the question of which axiom candidate is supported by better such arguments, we must face the prior question of whether these arguments carry any weight at all,

[113] See Chisholm (1977), 40.

and if so, why. We need to explain how, why, and to what extent such arguments count as evidence for the truth of their conclusions. Only then can we determine which among them constitute the better evidence.

A prerequisite for this inquiry is an appreciation for the rich variety of extrinsic supports offered by set theorists. The discussion in this chapter alone reveals a wide range of types:

1. Verifiable consequences. For example, Sierpiński's demonstration that many theorems proved from the axiom of choice can also be proved, though the proofs are more complicated, without it. Another case, to which attention was not explicitly drawn, provides support for measurable cardinals. Martin (1970) derived the determinacy of Borel sets (indeed Σ^1_1 sets) from the existence of a measurable cardinal. Five years later, in his (1975), this result was 'verified', that is, proved from ZFC alone.

2. Powerful new methods for solving pre-existing open problems. The axiom of choice settled the open question of whether or not the reals could be well-ordered. V = L and SC both provide methods for solving the long-standing open problems in analysis, and V = L even decides the continuum hypothesis.

3. Simplifying and systematizing theory. The axiom of choice brings order into the chaos of transfinite arithmetic.

4. Implying previous conjectures. The existence of a measurable cardinal implies that $V \neq L$, and a supercompact rules out projective well-orderings of the reals. Both these had been previously conjectured.

5. Implying 'natural' results. The zigzag pattern of separation properties in the projective hierarchy generated by projective determinacy (and hence by SC) is considered more natural than the Π-side pattern of V = L.

6. Strong intertheoretic connections. The detailed intertheoretic connections between determinacy and measurable cardinals, beyond Martin's result on Σ^1_1 determinacy, are too complicated to summarize here, but a simpler sort of example involves the extension of known patterns from one theory to the next. For example, in ZFC, every uncountable Σ^1_1 set has a perfect subset; in the presence of a measurable cardinal, this extends to Σ^1_2 sets, and with projective determinacy, to Σ^1_3 sets and beyond. Thus the three theories seem to be pulling in the same direction.

7. Providing new insight into old theorems. Projective determinacy allows many classical properties of Π^1_1 sets to be extended to Π^1_3. Along the way, this procedure often provides a new and simpler proof from ZFC of the classical theorem for Π^1_1.

The range of rules of thumb is hardly less bewildering than this array of extrinsic justifications. I've mentioned a handful marshalled in favour of large cardinals, but there are many more.[114] And even this rough-and-ready classification neglects the role of conjectures—like the non-existence of projective well-orderings of the reals—and other judgements of plausibility—like that against choice-generated oddities at low levels of the projective hierarchy. So even as description, leaving aside explanation, my account is far from complete. I'm sorry to say that I won't complete it here; filling in the details of structure of non-demonstrative, non-intuitive arguments and evaluating their cogency is a subject for another book, a book I unfortunately don't know how to write. After drawing attention to the problem, for now I can do little more than highlight its importance and encourage a concerted investigation. Let me conclude with a few words about how this last might best proceed.

There is an obvious similarity between this project and the central business of philosophers of science: giving an account of the confirmation of scientific theories. Indeed the very descriptions of the styles of extrinsic justification—verifiable consequences, simplifying and systematizing theory, strong intertheoretic connections—suggest that the analogy is a powerful one. It would seem that the compromise Platonist's science/mathematics analogy stands to gain further detail from these distinctive parallels between scientific and mathematical modes of theoretical justification.

To this line of thought, some will reply that the analogy is superficial, that natural scientists, not mathematicians, test their theories by experiment. The immediate response is that mathematicians *do* use experiments. Let me quote from Martin:[115]

I think that there has been more of what I might call subjecting determinacy hypotheses [to] experiment than is suggested by what I and other writers have said in print. An example: the first theorem in

[114] See Maddy (1988*a*).

[115] Personal communication, Sept. 1984.

determinacy I proved was that [the full axiom of determinacy] implies that every set of degrees of unsolvability contains or is disjoint from what is now called a 'cone'.[116] When I discovered the two-line proof of this, I was very excited. I was sure that, with a few minutes' thought, I would be able to find a set of degrees which was a counterexample. Thus I would refute [the full axiom of determinacy] and surely even [projective determinacy], and probably even Borel determinacy. I started going through various simple sets of degrees I knew about, checking each one out. I was surprised to discover that I could always find—by elementary recursion-theoretic means—the cone whose existence determinacy predicted . . . the effect on me was much as that on a physicist when a theory predicts a new kind of particle and such particles are then observed.

Here a mathematical experiment is undertaken quite explicitly. Other cases can be culled from material earlier in this chapter; for example, investigation of the consequences of V = L and SC for low-level projective sets can be viewed as tests.

But, opponents of the analogy might continue, these are experiments of a very different sort from those found in physics; no accelerators are involved, no observations of instruments or computer outputs, no cloud chambers, etc. Of course, this can hardly be denied, but what needs to be appreciated here is the extent to which scientific methodology varies even from one natural science to another. Martin's experiments may use paper and pencil and depend on previous results in recursion theory rather than using an electron microscope and depending on previous results in subatomic physics, but the botanist's experiments are different from both: in another era, she took a field trip and brought back hand drawings for comparison with previously gathered and classified samples. We don't expect a study of the methodology appropriate to physics to tell us all we want to know about how botanists, biologists, psychologists, astronomers, or geologists formulate and test their theories, so why should we expect mathematical science to conform to confirmation techniques drawn from some other science? The answer is that we shouldn't. If Martin's experiments are different from the physicist's, this should come as no surprise and shouldn't (by itself) count against their efficacy.

Respect for the variation between the sciences also undercuts the opposite, overly quick, reaction to the analogy between mathematics and natural science. Some, citing the similarities noted above,

[116] This theorem is the crucial lemma in Martin (1968).

might be inclined to conclude that the epistemological issues for this aspect of mathematical knowledge can be identified with the corresponding problems in the philosophy of science in general. The motive for such a move would be much like that of the logicists decades earlier; the problem of mathematical epistemology is reduced to another—the epistemology of logic or the epistemology of science in general—which is presumed to be an easier target. What I've been suggesting is that this won't work either, that the idiosyncrasies of mathematical theorizing require individual attention. A complete theory of the methods of physics (or psychology or biology), even if there were such a thing, would not be enough.

So the theory of mathematical theory formation and confirmation we're after will exploit parallels with the various natural sciences while attending to the unique aspects of mathematical methodology. And, if it is to do the ambitious job set for it here, it cannot rest content with pure description. When science is functioning smoothly, it may be enough to describe its methods, but in the case of contemporary set theory, even the practitioners aren't sure how the various non-demonstrative arguments should be evaluated. Thus Gödel writes that the recalcitrance of the continuum problem

> may be due to purely mathematical difficulties; it seems, however . . . that there are also deeper reasons involved and that a complete solution . . . can be obtained only by a more profound analysis (than mathematics is accustomed to giving) of the meanings of the terms . . . and of the axioms underlying their use. (Gödel (1947/64), 473)

I suggest that the 'more profound analysis' required is the very investigation I'm urging here. What's needed is not just a description of non-demonstrative arguments, but an account of why and when they are reliable, an account that should help set theorists make a rational choice between competing axiom candidates.

Finally, this fundamental problem, the problem of describing, explaining, and evaluating non-demonstrative arguments in mathematics, is a central challenge for the set theoretic realist, but it is not hers alone. Of course, any compromise Platonist will face it, but so will others. Anyone who holds that there is (most likely) a fact of the matter about the size of the continuum (or the open questions of Luzin and Suslin) must admit that one or the other of $V = L$ and SC is false. This leaves the problem of adjudicating between these two theories, and unless some alterative means is found, this in turn requires confronting and evaluating the non-demonstrative arguments for and against them. Actually, the range of Platonist and anti-Platonist philosophical positions for which this challenge remains a real one is quite broad, as will become clearer in my final chapter, but for all that, it has been almost universally ignored. The goal of this chapter will have been served if my portrait of the challenge itself is vivid and compelling enough that the reader now sees it as her own.

5
MONISM AND BEYOND

The main outlines of set theoretic realism are now in place. Its epistemology divides loosely into two parts, as befits a version of compromise Platonism: the intuitive (Chapter 2) and the theoretical (Chapter 4). Benacerraf's ontological puzzle is met by agreeing that numbers aren't objects, but insisting none the less on a close connection between numbers and sets, namely, that numbers are properties of sets (Chapter 3). Finally, though it is far from solved, the most serious open problem has at least been formulated with some care (Chapter 4).

Only a few bits of tidying-up remain for this closing chapter. First comes a final look at the ontology of set theoretic realism, this time with closer attention to the relationship between the mathematical and the physical. This is followed by two sections comparing and contrasting set theoretic realism with other contemporary positions, with an eye to illuminating some surprising convergences. I conclude, in section 4, with a summary and a look to the future.

1. Monism

Let me begin with a forceful objection Chihara once aimed at the very heart of set theoretic realism, that is, at my claim that 'sets of physical objects . . . have location in time and space and can be literally perceived by the senses'.[1] In the course of lampooning this position, he writes:

> Imagine that I am sitting at my desk. Its surface has been cleared of everything except an apple. Now according to Maddy, we can literally perceive on the desk, in addition to the apple, the set of apples on my desk (which happens to be a unit set). It is claimed that this set has a location in space, the exact spot where the apple is. Supposedly, it also came into

[1] Chihara (1982), 223.

existence at a particular time (when the apple did), and will go out of existence at a particular time (when the apple does). Obviously, Maddy thinks this set can be moved about in space. Now if we can perceive this set with the sense of sight, then what does it look like? Evidently, it looks exactly like the apple itself. After all, I don't see anything at that exact region in space that looks different from the apple. One wonders how this object is to be distinguished (perceptually) from the apple, since it has exactly the same shape and color. Perhaps it feels different. Let's touch it. But I can't feel anything there other than the apple. Evidently, this strange entity feels no different from the apple. How about its smell or taste? Again, it would seem that the set must be identical in smell and taste to the apple. So it looks, feels, smells, and tastes exactly like the apple and is located in exactly the same spot and at exactly the same time—yet it is a distinct entity! One would think that an entity with these properties would be of interest to the physicist. Furthermore, essentially the same reasons Maddy gives for maintaining that these sets can be perceived by the senses can also be given for claiming that a set of such sets can be perceived. Thus, we should be able to see the set whose only member is the unit set described above; we should be able to perceive the unordered pair consisting of the apple and the above unit set, and so on indefinitely. Presumably, all these different entities would look, feel, smell, and taste exactly alike. (Chihara (1982), 223–4)

Chihara's only explicit conclusion is the undoubtedly sound one that my view differs from Gödel's,[2] but the sense that this passage raises an important question for the set theoretic realist can hardly be avoided.

On the face of it, the question is one already considered in the second section of Chapter 2: how is it, on a given occasion, that we see a physical mass rather than a set, or one set rather than another, when all produce the same retinal stimulation? The answer there was that differences in training, or interests, or attention, could produce different cell-assemblies in different individuals, or facilitate the activation of one cell-assembly rather than another within a single individual. And the activation of different cell-assemblies, even given the same retinal stimulation, produces a difference in the purely phenomenological look of the scene.

In Chihara's case, then, the difference between the physical mass that makes up the apple and the singleton containing the apple is that the latter has an unambiguous number property—one—while the former is one apple, many cells, more molecules, even more

[2] See ch. 2, sect, 4, above.

atoms, and so on. What makes the example unsettling is that in this case, the singleton is so conspicuous that we rarely see the physical mass or any of the other sets. A topologist, heroically immersed in her work, might see left and right apple-halves, but it seems unlikely that even an infant would see a physical mass undifferentiated as a unit from its background—that is, something with no number property at all—and the normal adult almost invariably sees the single apple. Indeed the very question of how the physical mass differs from the singleton can be asked in such a way as to beg it: what's the difference between a single object and its unit set? A 'single object' already has an unambiguous number property! This doesn't mean that the physical mass doesn't differ from the singleton—it does—but once the physical mass is individuated, separated from its surroundings and seen as an isolated thing, that difference seems to evaporate.

So, while the set theoretic realist has a ready answer to one question—what distinguishes a physical mass from a set?—Chihara is asking another—what distinguishes an individuated physical object from its unit set? The answer to this new question cannot be that one has an unambiguous number property and the other doesn't, because both the single object and the singleton have the same number property: one. If there is a difference, it must lie somewhere else, and Chihara's remarks pointedly suggest that it is not perceptual.

The set theoretic realist's first option, in response to this situation, is to insist that there is an unperceivable difference. It wouldn't be the first such difference; we aren't very good at seeing the difference between gold and fool's gold, and that between water and heavy water is completely invisible. In science, unperceivable differences are detected by more sensitive instruments or implied by well-supported theory. Perhaps theoretical arguments could be found in mathematics, or more likely, in our theory of mathematics, in the philosophy of mathematics, for distinguishing individual things from their unit sets. These might have to do with inviolable differences between concrete and abstract, or between mathematical and physical.

The set theoretic realist's other option is simply to deny that there is any such difference at all, perceivable or otherwise, that is, to identify individuals with their singletons. This is not to say that every singleton is identical with its sole member; there is every reason to distinguish between $\{\{0, 1, 2, 3, \ldots\}\}$ and $\{0, 1, 2, 3, \ldots\}$, starting with the fact that one is finite and the other infinite. Rather, we take it that the physical objects, x, the individuals from which the generation of the iterative hierarchy begins, are such that $x = \{x\}$. After that, the axiom of extensionality[3] guarantees that sets formed at later stages will be distinct from their singletons. And, again, this option does not suggest that the physical mass of apple-stuff is identical with the singleton apple. Here there is a real difference in the determinacy of number property. All that's being denied is that the individual apple is distinct from its unit set.

I think both these options are open to the set theoretic realist, that is to say, both are fully consistent with the tenets of that position as described in previous chapters. And neither of them, as far as I can see, does any damage of the sort implied by Chihara's rhetoric. In particular, neither option gives up the claim to a real, perceptual difference between a set and an undifferentiated physical mass, between three apples and the unindividuated stuff that makes them up. Gödel, considering an argument from Russell, remarks:

> Russell adduces . . . against the extensional view of [sets] . . . the existence of . . . the unit [sets], which would have to be identical with their single elements. But it seems to me that these arguments could, if anything, at most prove that . . . the unit [sets] (as distinct from their only element) are fictions . . . not that all [sets] are fictions. (Gödel (1944), 459)

The same reply could be given to Chihara.

But even if both options are open, even if a set theoretic realist is free to follow either, my own preference is for the second. The only motivation I see for insisting on an unperceivable difference is the desire to maintain a strict dualism between the mathematical and the physical, and I feel no such desire. The remainder of this section will thus presuppose the identification of individual with singleton, but let me insist, one last time, that a set theoretic realist reluctant to take this turn should feel no obligation to do so.[4]

[3] Two sets are the same if and only if they have the same members. See Enderton (1977), 2, 17.

[4] Quine (1969*a*), §4) advocates the identification of object with singleton for the purpose of simplifying his formal theory. So doing requires a very minor modification of Zermelo–Fraenkel set theory and the additional assumption that the individuals, the elements from which the iterative hierarchy begins, form a set with at least two members. See my 'Physicalistic Platonism' (forthcoming), especially the appendix.

While Chihara's example is aimed at one specific aspect of set theoretic realism, other anti-Platonists may be bothered by a more general worry arising from versions of physicalism.[5] Physicalism began in the hands of the positivists as a very strong thesis about the reducibility of all sciences to the vocabulary of fundamental physics.[6] In this form, it foundered on the methodological independence of the special sciences,[7] but the crude idea that physics is somehow basic retains its appeal. A sophisticated version appears in the writings of the contemporary physicalist Hartry Field.

Field describes the doctrine as a well-supported methodological principle:

> . . . physicalism [is] the doctrine that chemical facts, biological facts, psychological facts, and semantical facts, are all explicable (in principle) in terms of physical facts. The doctrine of physicalism functions as a high-level empirical hypothesis, a hypothesis that no small number of experiments can force us to give up. It functions, in other words, in much the same way as the doctrine of mechanism (that all facts are explicable in terms of *mechanical* facts) once functioned . . . (Field (1972), 357)

Mechanism was rejected when Maxwell's theory of electromagnetism could not be explained in mechanical terms. Field concludes that

> Mechanism has been empirically refuted; its heir is physicalism, which allows as 'basic' not only facts about mechanics, but facts about other branches of physics as well. I believe that physicists a hundred years ago were justified in accepting mechanism, and that, similarly, physicalism should be accepted until we have convincing evidence that there is a realm of phenomena it leaves out of account. (Field (1972), 357)

He gives this example of how the physicalistic principle functions:

> Suppose, for instance, that a certain woman has two sons, one hemophilic and one not. Then, according to standard genetic accounts of hemophilia, the ovum from which one of these sons was produced must have contained a gene for hemophilia, and the ovum from which the other son was produced must not have contained such a gene. But now the doctrine of physicalism tells us that there must have been a *physical* difference between

[5] I mentioned this possibility in ch. 2, sect. 1, above but postponed discussion until now.

[6] See Carnap (1934).

[7] See e.g. Fodor (1975), 9–26.

> the two ova that explains why the first son had hemophilia and the second one didn't . . . We should not rest content with a special biological predicate 'has-a-hemophilic-gene'—rather we should look for non-biological facts (chemical facts; and ultimately, physical facts) that underlie the correct application of this predicate. (Field (1972), 358)

Field does not require that the biological predicate 'has a haemophilic gene' be translated into the basic vocabulary of physics or even chemistry, or that all biological laws be derivable from the laws of physics or chemistry, or that any rational biological methodology be identical with that of physics or chemistry. He only insists that there be a chemical and ultimately physical explanation of why this particular ovum has a haemophilia gene and that one doesn't.[8]

Now presumably a physical explanation is one that involves only physical things, physical facts. If physical things and physical facts are just those spoken about in physics, then the indispensability arguments make it hard to see why physicalism presents a problem for any version of Platonism; according to Quine and Putnam, physics speaks constantly and essentially of things mathematical. If being part of physics were all there were to being physicalistically acceptable, mathematical things would get into the physicalist's ontology on the ground floor, even ahead of the chemical, the geological, the astronomical, the biological, and so on.

Obviously, for those who take physicalism to raise a problem for Platonism, 'being physical' comes to more than 'being mentioned in physics'. To see what this something more might be, consider once again Field's version of the epistemological problem for Platonism:

> what raises the really serious epistemological problems is not merely the postulation of causally inaccessible entities; rather, it is the postulation of entities that are causally inaccessible *and* can't fall within our field of vision *and* do not bear any other physical relation to us that could possibly explain how we can have reliable information about them. (Field (1982), 69)

In contrasting mathematical entities with his own space-time regions, Field clarifies the 'physical relations' he has in mind:

> there are quite unproblematic physical relations, viz., spatial relations, between ourselves and space-time regions, and this gives us epistemological

[8] In his 'Physicalism' (forthcoming), Field differentiates his view from various stronger and weaker versions of physicalism.

access to space-time regions. For instance, because of their spatial relations to us, certain space-time regions can fall within our field of vision. (Field (1982), 68)

To be epistemologically accessible, to be acceptable on Field's world-view, an entity need not be causally efficacious, but it must at least be spatio-temporally located. Assuming that acceptable entities, for Field, are physical, this suggests that what it takes to be physical, over and above being talked about in physics, is location in space and time, and preferably in the causal nexus.[9]

On this reading, physicalism creates obvious problems for traditional Platonism, with its acausal, non-spatio-temporal objects. The set theoretic realist, by contrast, begins in a much stronger position: her sets of medium-sized physical objects can 'fall within our field of vision'. Of course, not all sets actually do fall within our field of vision, but, to paraphrase Field, 'this raises no more epistemological problems for [sets] than it raises for, say, tigers' (p. 68). If all that's required for physicalistic acceptability is spatio-temporal location, the set theoretic realist's impure sets are unimpeachable.

But what about pure sets? What about the empty set, the set of von Neumann ordinals, and its power set? Those unafflicted by physicalistic scruples are free to respond that we gain knowledge of the pure sets by theoretical inference from our elementary perceptual and intuitive knowledge of impure sets, but physicalists will complain that this reply does nothing to solve the problem of spatio-temporal location. Does this mean that such a physicalist must reject set theoretic realism after all? This follows only if set theoretic realism is irrevocably committed to pure sets, and I contend that it is not.

In fact, the pure sets aren't really needed. The set theoretic realist who would simultaneously embrace physicalism can take the subject matter of set theoretic science to be the radically impure hierarchy generated from the set of physical individuals by the usual power set operation, except that the empty set is omitted at each stage. On this picture, each set, no matter how exalted in rank, is located where the physical stuff in its transitive closure is located. The theory of this structure differs only trivially from that of the

[9] Cf. Armstrong (1977).

usual hierarchy with individuals.[10] It can serve all the same purposes.

So the set theoretic realist can locate all the sets she needs in space and time. Still, a more physicalistically satisfying ontology would be not only spatio-temporally located, but causally efficacious as well. But notice: if sets are indeed perceivable, as I've argued, then they must play the same role in the generation of my perceptual beliefs about them as, say, my hand plays in the generation of my perceptual belief that there is a hand before me when I look at it in good light, a role which is, presumably, causal. Or, to use a non-psychological example, suppose you deposit three quarters in a soft-drink machine and a soda drops out. Which properties of that which you deposited are causally responsible for the emergence of the Pepsi? Well, the weight of the physical mass of metal, its shape, and also the number property: three. (The machine counts somehow.) From the set theoretic realist's perspective, that which has a number property, that is to say, a set, *is* causally efficacious.

I conclude that set theoretic realism is consistent with physicalism. Once again, this is not to say that every set theoretic realist must be a physicalist; non-physicalists may prefer to retain the standard version of set theoretic realism with its pure sets. My point is that the position can be physicalized without significant trauma.

In this section, I've suggested two minor alterations in the set theoretic realist's ontology: the identification of physical objects with their singletons and the elimination of pure sets. Together, these moves produce a powerfully symbiotic picture of the relationship between the mathematical and the physical: every physical thing is already mathematical, and every mathematical thing is based in the physical. In place of the customary dualism of

[10] With two individuals, x and y, a version of the ordinals can be constructed without pure sets—x, $\{x, y\}$, $\{x, y, \{x, y\}\}$, and so on—and the various axioms and theorems can be tinkered with. Practically speaking, however, it is probably best to keep the empty set as a notational convenience. Gödel, for example, suggests it could be treated as a fiction 'introduced to simplify the calculus like points at infinity in geometry' (1944, p. 459). Fraenkel, Bar-Hillel, and Levy say the empty set is introduced for 'reasons of convenience and simplicity, and can be regarded as a mere notational convention' (1973, p. 24), and even Zermelo calls it 'fictitious' (1908*b*, p. 202). See my 'Physicalistic Platonism' (forthcoming), appendix.

mathematical and physical, this pared-down set theoretic realism offers a version of monism.[11]

To appreciate just how closely the two are intertwined on this view, try to separate them. A purely mathematical world would be empty. What would a purely physical world be like? As soon as there are number properties, there are sets that bear them, so a world without mathematical things would have to be a world without *any* things, a completely amorphous mass: the Blob. To add even the structuring into individual physical objects is to admit singletons, to broach the mathematical. The only way to confine ourselves to the purely physical is to refrain from any differentiation whatsoever.

Perhaps such a world is possible, but it clearly isn't our world, with its objects, kinds, patterns, and structures of so many, widely varied sorts. In place of the old picture—physical reality here and now, mathematical reality nowhere and nowhen—set theoretic monism offers a spatio-temporal reality inseparably physical *and* mathematical. Physics and mathematics, on this new picture, are two sciences, along with chemistry, biology, psychology, and the rest, that study aspects of this reality. Each science has its own vocabulary and laws, its own techniques and methods, but this doesn't mean that the world itself is divided into the physical, the mathematical, the chemical, the biological, the psychological, and so on. Rather, everything is ultimately physico-mathematical or mathematico-physical.

Some will note that, strictly speaking, this view is more Aristotelian than Platonistic. They are right, in the sense that Aristotle's forms depend on physical instantiations, while Plato's are transcendent. I retain the term 'Platonism' here, not for its allusion to Plato, but because it has become standard in the philosophy of mathematics for any position that includes the objective existence of mathematical entities. Others will point out that singletons don't deserve the special status my presentation has awarded them; for the monist, two objects (as opposed to the undifferentiated mass of physical stuff that makes them up) are already a doubleton, as well. This is also correct. The singleton case is unique only in its psychological impact: it makes us realize just how little it takes for mathematics to intrude. But for all this, I hope

[11] For those who think of Platonism as a form of theology, this is a version of pantheism.

the general features of set theoretic monism are clear enough. I leave it as an option for any set theoretic realist to whom it might appeal.

2. Field's nominalism

Despite my efforts to turn away objections to Platonism based on epistemology, on the possibility of multiple reductions, and on purely physicalistic concerns, Field remains unconvinced. Instead of seeking out new objections to counter, a more fruitful strategy at this stage might be to have a look at Field's nominalistic alternative to Platonism. Field is not the only nominalist on the contemporary scene, but his version has the distinction of being non-revisionist:[12] he sets out to show how we can go on using classical mathematics in science exactly as we have been, but without admitting the existence of mathematical things, and he attempts this without short-changing the Quine/Putnam indispensability arguments. Here is a form of nominalism that should give the Platonist pause.

Field and I agree that the indispensability arguments provide the best evidence for mathematics as a whole. Moving beyond Quine/Putnamism into Gödelian territory, we also agree that there are other possible forms of mathematical evidence:

> if we assume that there is at least one body of pure mathematical assertions that includes existential claims and that is true . . . then we are assuming that there are mathematical entities. From this we can conclude that there must be some body of facts about these entities, and that not all facts about these entities are likely to be relevant to known applications to the physical world; it is then plausible to argue that considerations other than applications to the physical world, for example, considerations of simplicity and coherence within mathematics, are grounds for accepting some proposed mathematics axioms as true and rejecting others as false. (Field (1980), 4)

Finally, we agree that the Benacerraf-style worry is a real one, that the Platonist owes a descriptive and explanatory account of our knowledge of (or reliability about) mathematical facts, and I can

[12] Recall from ch. 1, sect. 4, that the nominalist claims there are no mathematical entities. Chihara (1973, ch. 5) is also a nominalist, but he proposes an abbreviated and reinterpreted mathematics that may or may not be adequate for scientific purposes.

extend this accord by accepting for now the further stipulation that this account must satisfy the physicalist.

Against this shared backdrop, two possible strategies stand out: take mathematical statements to be (mostly) true and meet the epistemological challenge head on, or take mathematical statements (at least the existential ones) to be false and explain why these falsehoods are so useful in applications. Field chooses the second, which requires him to somehow circumvent the indispensability arguments. What I want to suggest here is that this arduous undertaking does not win him the advantages he hopes for, indeed that it doesn't exempt him from what I take to be the most serious challenge to Platonism.

The best approach to Field's methods is to return to the thinking of the traditional Platonist: mathematics, if true, is necessarily true, that is, it is true regardless of the contingent details of the physical world, true in all possible worlds. Let M be our necessarily true mathematical theory. Now suppose that N is a consistent theory of what the physical world might be like, and further suppose that N is nominalistic, that it makes no reference to mathematical entities. Then N + M must also be consistent; otherwise, the truth of N would imply the falsehood of M, and M couldn't be true in a possible world where N is true, contradicting the necessity of M. So, if M is our true mathematical theory, it must be consistent with any consistent nominalistic theory N.

Now suppose that A is a nominalistic statement implied by N + M. Then N + not-A + M is inconsistent. But we've just finished arguing that M must be consistent with any nominalistic theory that is itself consistent, so it follows that N + not-A is also inconsistent. By elementary logic, this means that N implies A. So we've shown, from the perspective of the traditional Platonist, that if a nominalistic statement follows from a nominalistic theory plus mathematics, then that same nominalistic statement follows from the nominalistic theory alone. In technical terms, mathematics is conservative over nominalistic physical science.[13]

[13] The traditional Platonist's argument doesn't work for the set theoretic realist because she doesn't take mathematics to be necessary (see ch. 2, sect. 2, above). The set theoretic monist also denies that there is such a thing as nominalized physics, because physical objects are already mathematical, which severely undercuts the potential significance of any conservativeness claim.

Of course, as a nominalist, Field rejects the assumption that mathematics is true at all, let alone necessarily true, but he agrees with the conclusion that good mathematical theories should be conservative:

> it would be extremely surprising if it were to be discovered that standard mathematics implied that there are at least 10^6 non-mathematical objects in the universe, or that the Paris Commune was defeated; and were such a discovery to be made, all but the most unregenerate rationalist would take this as showing that standard mathematics needed revision. *Good* mathematics *is* conservative; a discovery that accepted mathematics isn't conservative would be a discovery that it isn't good. (Field (1980), 13)

This conservativeness is a boon to the nominalist:

> even someone who doesn't believe in mathematical entities is free to use mathematical existence-assertions in a certain limited context: he can use them freely in deducing nominalistically-stated consequences from nominalistically-stated premises. And he can do this not because he thinks those intervening premises are true, but because he knows that they preserve truth among nominalistically-stated claims. (Field (1980), 14)

So this is the beginning of Field's answer to the indispensability argument: he admits that mathematics is used in science to derive physical claims from other physical claims, but insists that we can believe these results without believing the mathematics employed to be true. We need only believe it is conservative, that whatever it implies is already implied by the physical theory itself.

But this is only part of the story, for the role of mathematics in deriving one physical statement from another is only one of its roles in physical science. As Putnam has argued, many of the physical statements themselves make essential appeal to mathematical entities. In other words, the nominalistically stated physical claims discussed so far don't cover most of physical science. Field is fully sensitive to this point. He concludes only that '*once such a nominalistic axiom system is available*, the nominalist is free to use any mathematics he likes for deducing consequences, as long as the mathematics he uses [is conservative]' (Field (1980), 14). So the answer to the indispensability arguments comes in two parts. First it must be shown that physical theories can be stated without the use of mathematics, then classical mathematics must be shown to be conservative over those restated physical theories. That accomplished, the scientist can use whatever mathematics she likes in deriving nominalistic consequences from nominalistic theory, because any such consequence derived using mathematics is already implied by the nominalistic theory alone.

For concreteness, let me rehearse a simplified but I hope illustrative example of this strategy. We ordinarily use real numbers to measure distances in our theory of space. Thus, by a standard indispensability argument, any confirmation of our theory of space also confirms the existence of real numbers. But, Field argues, the use of the reals in this context is actually dispensable after all. To show this, he must first reformulate our standard theory of space without talk of distances. There is a way to do this, codified by Hilbert;[14] rather than assigning locations and distances to points, it makes use of comparative predicates like 'between' and 'congruent'. This theory is nominalistic because it deals only with points and regions of space and not with numbers. Call it H.

Now suppose we'd like to establish some nominalistic claim A about space. To apply real number theory, we first move to a larger theory, S, that combines H with some set theory. In S, we can prove that there is a function from pairs of points to real numbers[15] that does all the right things: for example, the segment between x and y is congruent to the segment between x' and y' in H if and only if the real number assigned to (x, y) is the same as the real number assigned to (x', y'). In fact, we can show that the space itself is isomorphic to the set of ordered triples of real numbers. In this rich context, we translate A into an equivalent statement A$'$ that talks about distance and hence about real numbers, and we proceed to prove A$'$. Because A and A$'$ are equivalent, this also establishes A in the theory S. But S, being a good mathematical theory, is conservative over H, so A is also implied by H alone. And that was what we wanted to show in the first place. But we needn't assume the truth of S to do it, only its conservativeness.[16]

Field extends this technique to cover applications of classical

[14] See Hilbert (1899).

[15] Whichever version of the reals we select for this measuring job.

[16] Students of Field's theory will realize that I'm confining my attention to the second-order version of his view. (See Shapiro (1983*b*) and Field (1985) for discussion of this distinction.) I do this because it is the second-order version that offers the full use of classical mathematics without an ontology of mathematical entities—the first-order version offers something less—and because I think our understanding of space (and other mathematical notions) is essentially second-order. (See Shapiro (1985) on this last point.) For an introductory account of the differences between first- and second-order logic, see Enderton (1972), ch. 4.

mathematics in Newton's gravitational theory, but some commentators[17] doubt that it can be adapted to other parts of physics, in particular to quantum mechanics. Because his efforts to date are (at best) only partial, Field admits that the indispensability arguments retain some force:

> At present of course we do not know in detail how to eliminate mathematical entities from every scientific explanation we accept; consequently, I think that our inductive methodology does at present give us some justification for believing in mathematical entities. But . . . justification is not an all or nothing affair. . . . what we must do is make a bet on how best to achieve a satisfactory overall view of the place of mathematics in the world. . . . my tentative bet is that we would do better to try to show that the explanatory role of mathematical entities is not what it superficially appears to be; and the most convincing way to do that would be to show that there are some fairly general strategies that can be employed to purge theories of all reference to mathematical entities. (Field (1989), 17–18)

Weighing what he sees as the epistemological and ontological drawbacks of Platonism against the indispensability arguments, Field wagers that mathematics can be shown to be dispensable, after all. This is his project.

When this version of nominalism is compared with traditional Platonism, some observers[18] argue that its space-time points and regions are abstract, and thus as susceptible as numbers to epistemological challenge. From the physicalistic point of view sketched in the last section, this can hardly be true. Space-time points and regions have location, and some such regions 'fall within our field of vision'. As I've indicated, the same is true of the set theoretic monist's impure sets, so at this crude level, nominalism and monism are on a par, and both are preferable to traditional Platonism.

Like the traditional Platonist, Field is also faced with the accusation that his entities are causally inert. The argument runs that it is the objects in space-time, not space-time itself, that are causally efficacious. Field responds to this charge along two different lines. First, he suggests that physical objects be identified

[17] e.g. Malament (1982).
[18] e.g. Resnik (1985*a*).

with the space-time regions they occupy.[19] In this way, at least the occupied areas of space-time enter into the causal nexus. A stronger argument involves the claim that

> a field theory is *most naturally* construed as a theory that ascribes causal properties . . . to space-time points. (Field (1982), 70)

> In electromagnetic theory for instance, the behavior of matter is causally explained by the electromagnetic field values at unoccupied regions of space-time . . . (Field (1980), 114)

Given the omnipresence of fields, this observation brings causal powers to all points of space-time. So once again, the nominalist and the monist are on equal footing.

To further the comparison, we must consider the respective ontologies more closely. The nominalist's world consists of space-time regions; ordinary and theoretical physical objects are identified with the space-time regions they occupy, and points can be identified with partless regions. The monist's world consists of sets; physical objects are singletons among these. The contrast between the two views becomes smaller when we realize that a thorough account of the monist's discrete physical objects will involve the study of perceptual continua as well: object boundaries, trajectories of movement, etc.[20] Thus spatial point sets join the monist's ontology. So both nominalism and monism embrace discrete objects and continua: for the former, both are species of space-time regions; for the latter, both are species of sets.

Finally, what of that old point of contention, the numbers? The nominalist, of course, eschews them. For the monist, as we've seen,[21] the question of the existence of numbers is a special case of the age-old problem of the existence of universals. Now Field sides with old-fashioned nominalism against universals as well as his modern variety against mathematical entities,[22] and I see no reason why the monist's commitment to numbers need be any stronger than Field's commitment to the properties of his regions. In so far as old-fashioned nominalism is tenable, the monist can agree with Field that numbers don't exist.

So far, then, the nominalist and the monist are not as far apart as

[19] Field (1982), 70.
[20] See ch. 2, sect. 3, above.
[21] In ch. 3, sect. 2, above.
[22] See Field (1980), 35, 55–6, and (1982), 70.

rhetoric would suggest—where the nominalist sees a space-time region containing the stuff of three apples, the monist sees a set of three apples—but a dramatic difference soon emerges. These space-time regions are the end of the ontological story for the nominalist, but the monist's world, in addition to sets of apples, also contains sets of sets of apples, sets of sets of sets of apples, and so on. Physicalistically speaking, these sets of higher rank are no clear liability; they have location and they can (at least in principle) be causally efficacious. If physicalism doesn't rule them out, we should ask what motivates the monist to include them and what the nominalist hopes to gain by abstaining.

Part of the motivation for an escalation of ranks lies in arithmetic itself. Two pairs of shoes are naturally viewed as a set of two sets; a series of sets of ever-increasing rank, analogous to the von Neumann ordinals, does good service as a measuring device for the number properties of finite sets; this set theoretic context allows us to prove the Peano axioms and to provide a simple and explanatory theory that encompasses and explains various well-entrenched generalizations, for example that the union of two disjoint two-membered sets has four members. By contrast, the nominalist's position here is much like that of the aggregate theorist considered back in Chapter 2: a statement of number concerns a mass of physical stuff together with a predicate. On the one hand, it isn't clear that a smooth and flexible arithmetic can be established on this basis, but, on the other, I doubt that a truly persuasive case of the postulation of infinite ranks can be based on arithmetic alone.[23]

Let me turn, then, to the continua which inhabit both the nominalist's and the monist's universes. Both theorists hypothesize that these satisfy the nominalist's Hilbert-style axioms, and, given that they are physical entities, the nominalist should be as interested as the monist in answering further questions about them.[24] A number of deep questions can be asked about physical structures of this complexity,[25] and it is the consequences of asking them that I now want to explore.

[23] I consider this possibility at greater length, but no more conclusively, in my 'Physicalistic Platonism' (forthcoming), § 7.

[24] Field's remarks about mathematics quoted at the beginning of this section strongly suggest that as soon as entities are admitted into the nominalist's ontology, all facts about them are worthy of investigation.

[25] Many observers have remarked that these questions include the continuum hypothesis. See Resnik (1985*b*), 198. I'll consider other questions here, for reasons that will become obvious.

Recall (from section 4 of Chapter 4) that being determined is a property of sets of reals. This property can be defined mathematically in fairly simple terms; its nominalistic counterpart should be expressible in Field's nominalistic theory.[26] Now suppose that our nominalist observes that a great number of the simple regions he comes across are in fact determined. As long as he considers nothing more complicated than the regions corresponding to what the monist would think of as countable unions and intersections generated from open sets,[27] he will encounter no set that is not determined. As a scientifically minded inquirer, he will want an explanation of this fact.

The monist has an explanation for the corresponding fact about the world of sets in the form of Martin's theorem that all Borel sets are determined.[28] This explanation involves, however, an inescapable commitment to infinite ranks; Harvey Friedman has shown that the theorem requires them.[29] So we imagine our scientifically minded monist hypothesizing the axioms of infinity and replacement, expanding her ontology accordingly, in order to gain an explanatory theory of the behaviour of Borel sets.[30]

Where does this leave our nominalist? Presumably, he'd also like to explain the behaviour of his Borel regions; surely he wants a theory of his continuum that explains as much as the monist's. It might be thought that the nominalist will have to break down and postulate some aggregate-theoretic analogue of the monist's higher ranks in order to gain a counterpart to the theorem on Borel regions. But to think this is to forget the role of conservativeness. Recall that if ZFC, which includes replacement, is conservative over the nominalist's theory H, then whatever ZFC can prove is already true in the nominalist's world. In particular, if ZFC is conservative over H, then all Borel regions are determined, regardless of whether or not there really are higher ranks of any kind.

So, in order to show that Borel regions are determined, the

[26] Saying a set is determined involves quantification over real numbers. Methods of Shapiro (1983*b*) and Resnik (1985*b*) show how to simulate such quantification in Field's system using space-time points.

[27] Field (1980, p. 63) describes the nominalistic version of open sets. Countable unions and intersections generate the Borel sets.

[28] Martin (1975; 1985).

[29] Friedman (1971).

[30] Of course, this isn't the only reason for assuming infinity or replacement, but I'm simplifying here. For more, see Maddy (1988*a*), §1.8.

nominalist need only establish the conservativeness of ZFC over the theory H. In his book, Field gives a set theoretic proof of this fact in a theory slightly stronger than ZFC itself, but this is a metamathematical argument in terms of models that is obviously unavailable to the nominalist. In fact, conservativeness itself is usually defined metamathematically,[31] so it is unclear that there is even a legitimate nominalistic version of the bare claim that ZFC is conservative.

Field's reply is that the conservativeness of ZFC, nominalistically stated, comes to a claim about logical possibility, where logical possibility is a primitive notion not defined, as it is classically, in metalogical terms. What the nominalist needs to know, then, is (slightly more than)[32] that ZFC is logically possible in this sense. How does he know this? Well, presumably the monist also claims to know it, or something very like it, and Field suggests that

> it is no more problematic for a nominalist to claim that a belief that [ZFC is logically possible] is reasonable than it is for a platonist to make this claim. (Field (1985), 140)

Thus, for example, both the nominalist and the monist can cite the fact that no one has yet derived a contradiction from these axioms.[33]

This situation can be clarified if we consider one more hypothetical example, returning this time to the concerns of the classical analysts. Let's see what happens if the nominalist examines his regions and asks his nominalistic version of the question: is this region measurable?[34] Once again, the simplest regions he considers will be measurable. Indeed, this time he can go beyond the Borel regions to the Σ^1_1 regions and still not encounter anything non-measurable. But what will happen if he goes further? The complements of the Σ^1_1 regions, the Π^1_1 regions, will still be measurable, but what about their projections, the Σ^1_2 regions, and beyond?

Of course, the monist has considered these questions in Platonistic terms, and we've seen that her theory can answer them

[31] M is conservative over N if and only if, for any nominalistic assertion A, if A is true in all models of M, then A is true in all models of N.

[32] See Field (1985), 139–40.

[33] See Field (1984), 88, 124.

[34] Field mentions the possibility of non-measurable regions (1980, p. 144, n. 26). And again, measurability is a property that can be stated mathematically using only quantification over reals, and thus, should be expressible in Field's system.

definitively.[35] What makes this case different from the last is that this time the monist has an embarrassment of riches: two competing theories that answer these questions in different ways; V = L answers that there is a Δ^1_2 non-measurable set, while SC implies that all projective sets are measurable. In other words, as we've seen, the monist is placed in a position analogous to that of the natural scientist who must chose between two competing theories, and the reaction of the set theoretic community is much as this analogy would suggest: theorists disagree, they offer competing evidence, and further evidence is sought to decide the matter. I've suggested that describing and explaining the justificatory power of this sort of evidence is the main open problem for contemporary Platonism.

Where does this leave the nominalist? In order to see which of his regions is measurable, he needn't worry over which of the platonist's two opposing theories is true; he need only know which one is conservative.[36] In this connection, Field says much what he did before:[37]

> any reason that a platonist offers for believing that it is [ZFC + SC] rather than [ZFC + V = L] that is true [or vice versa] can be taken over by a nominalist to argue with just as much force that it is [ZFC + SC] rather than [ZFC + V = L] that is possible. (Field (1985), 140)

In other words, the nominalist is faced with exactly the same bewildering array of argument and counter-argument, evidence and counter-evidence, as the monist. Which means that in order to answer his physical question about the measurability of his regions, the nominalist must face a carbon copy of the most difficult epistemological open question that confronts the monist: what makes these arguments good or bad? Thus, despite his noble effort to refrain from committing himself to higher ranks, or anything like them, the nominalist has not thereby saved himself from the monist's most difficult epistemological challenge.

[35] See ch. 4, sect. 4, above.

[36] I say 'which one' because at most one can be conservative. This is because part of conservativeness is second-order semantic consistency, and all standard models of set theory agree on the structure of the reals and their subsets. Of course, both theories might fail to be conservative.

[37] Field actually speaks, not of ZFC + V = L and ZFC + SC, but in general, of two set theories M and M* which are related as these two are.

Two conclusions follow. The first involves Field's wager that dispensing with mathematics in science will produce a better overall theory than Platonism can provide. I've argued that physicalism gives us no reason to prefer Field's nominalism to the monistic version of set theoretic realism. Further, the elementary epistemology of monism and this nominalism are comparable; at the most fundamental level, sets or regions 'fall within our field of vision'. Finally, both the monist and the nominalist are faced with a difficult epistemological question at the theoretical level. I conclude that Platonism need be no more problematic than Field's nominalism, and thus, that there is nothing to balance the disadvantages of nominalism—e.g. the need to rewrite science—when the overall merits of the two theories are compared. On these terms, monism should prevail.

Before turning to my second conclusion, let me pause to draw a moral from this one. We've seen that the mathematical theory of sets has its metaphysical as well as its historical roots in the theory of the continuum, in the calculus and higher analysis. The Platonist takes the ontology of this theory at face value. The nominalist tries to abstain from this as a way of avoiding some difficult philosophical questions. I think the moral of this discussion is that anyone, including the nominalist, who embraces the full continuum and analysis in some form or another, will end up, sooner or later, meeting those same difficult philosophical problems, perhaps lightly disguised, but stubbornly undiminished. This lesson falls neatly in line with the realistic urgings of Frege, Quine, and Putnam.[38]

My second conclusion transcends the pro-Platonist propaganda that is the main theme of this book. Towards the end of Chapter 4, I suggested that set theoretic realism in particular, and compromise Platonism in general, are not the only positions faced with the difficult problem of assessing and explaining the rationality of the non-demonstrative arguments for and against $V = L$ and SC. The discussion in this section shows, quite surprisingly, that Field's nominalism provides a fresh example; in order to answer what he counts as physical questions, questions about space-time-regions, he must deal with this same open problem. A problem common to nominalist and Platonist is likely to be an extremely fundamental

[38] See ch. 1, sect. 4, above.

one, deserving the attention of philosophers from a wide range of persuasions. We'll meet with two more in the next section.

3. Structuralism

Before closing, I'd like to touch briefly on one other conspicuous position in the philosophy of mathematics of recent years, namely structuralism. While lip-service to the general idea behind this view is fairly common—mathematics is about structures, not objects—there is no complete and definitive statement of structuralist orthodoxy comparable to Field's writings on his own version of nominalism. Michael Resnik and Stewart Shapiro offer the most comprehensive contemporary statements,[39] but this work is still in progress, so a thorough comparison between structuralism and set theoretic realism will have to wait till another day. My more modest goal here is simply to sketch the main outlines of a structural approach to mathematics and to suggest that it differs less from set theoretic realism than partisan rhetoric would indicate.

Though structuralist thinking goes back at least to Dedekind,[40] modern versions are inspired by considerations akin to those in Chapter 3 above. The Platonist claims that mathematics is about objects, but, as Benacerrafian meditation on multiple reductions indicates, we seem to know nothing about these objects other than that they are related to one another in certain ways. If mathematical objects have distinguishing features over and above these, those properties are hidden and presumably unimportant to the mathematician. How, for example, are we to say which particular objects are the natural numbers?

[39] See Resnik (1975; 1981; 1982), Shapiro (1983*a*; 1985; forthcoming). (Parsons (forthcoming) is another valuable resource.) One major difference between these two is Resnik's preference for first-order formulations, e.g. number theory is the study of any structure—there are many—that satisfies the first-order Peano axioms. This smacks to me of if-thenism. Shapiro, by contrast, thinks of arithmetic as the study of a single particular structure, the one specified by the second-order axioms. Here I side with Shapiro. My thanks go to both writers for their help with this section.

[40] See Parsons (forthcoming), § 3. On the recurrence of structuralism, Parsons remarks, 'its tendency to be revived after attempts run into serious difficulties shows that the ontological intuition behind it exerts a powerful attraction' (p. 8).

The structuralist replies by rejecting the presuppositions of the question:

In mathematics . . . we do not have objects with an 'internal' composition arranged in structures, we have only structures. The objects of mathematics, that is, the entities which our mathematical constants and quantifiers denote, are structureless points or positions in structures. As positions in structures, they have no identity or features outside of a structure. (Resnik (1981), 530. See also Shapiro (1983*a*), 534)

Arithmetic, for example, is not the study of certain objects, the numbers, but the study of the natural number structure, an endless sequence of featureless positions satisfying certain conditions. One instantiation of that structure is the von Neumann ordinals, another is the Zermelo ordinals. But sets themselves are also positions in a structure, so the multiple reductions of number theory to set theory just show that the natural number structure occurs many times within the set theoretic hierarchy structure.

Some structures are physically instantiated: for example, the substructure of the natural number structure consisting of its first three positions is instantiated by the apples on the table. On the other hand, many patterns of higher mathematics—e.g. the iterative hierarchy structure—presumably are not physically realized. Between these extremes, some applications of mathematics in science come to the postulation of enough theoretical physical entities to exemplify the relevant structure:

the claim that actual space exemplifies the structure of Euclidean geometry involves an assertion that there is a continuum of space points. . . . science . . . proceed[s] by discovering mathematical structures exemplified in material reality, but the discovery is often indirect and involves the postulation of theoretical entities. (Shapiro (1983*a*), 540)

In sum: universals are the subject matter of mathematics; some but not all of these universals are physically instantiated.

Already, certain points of contact between structuralism and set theoretical realism are obvious: both solve the multiple reductions problem by exchanging objects for universals.[41] Indeed much of their talk is strikingly similar. In the three-apple case, the set

[41] The structures themselves are universals (Shapiro (1983*a*), 536) or, in Resnik's terminology, 'patterns'. The positions in these structures count as 'objects' for both, but notice that these objects are structure-dependent: they have no properties apart from the relations they bear to other positions in the same structure.

theoretic realist says there is a set on the table that is equinumerous with $\{\phi, \{\phi\}, \{\phi, \{\phi\}\}\}$. The structuralist says there is a physical arrangement on the table that instantiates the same pattern as $\{\phi, \{\phi\}, \{\phi, \{\phi\}\}\}$ under the successor relation. Both are claiming that a physical mass has a certain organization. One calls that organization forming a set equinumerous with $\{\phi, \{\phi\}, \{\phi, \{\phi\}\}\}$, the other calls it instantiating the same pattern as $\{\phi, \{\phi\}, \{\phi, \{\phi\}\}\}$. At this point, I think it's fair to wonder if any real significance attaches to this difference in description,[42] but I won't undertake to answer that question here.

There are also agreements in epistemological thinking. For the structuralist, various claims about a pattern of dots on a piece of paper

> are simply obvious to anyone who has sufficient mathematical experience to understand them and who attends to the diagram. . . . they are *in a sense* read off the drawing. So long as we are taking our perceptual faculties for granted, they need no further justification. . . . [they] continue to hold when talk of dots is replaced by talk of a sequence of squares, stars, a row of houses, a stack of coins, etc. . . . These additional assertions are as evident or almost as evident as the original ones. We have thus arrived at knowledge of an abstract *pattern* or *structure*. (Resnik (1975), 34)

In place of the non-spatio-temporal, causally inert mathematical entities of traditional Platonism, the structuralist substitutes perceivable arrangements of things.[43] The existence of infinite patterns and facts about them are then justified theoretically:[44]

> If [our theory of the infinite structure] turns out to be highly coherent and confirmed by our knowledge of the finite patterns from which it arose, then our belief in the existence of the pattern is justified. (Resnik (1975), 36–7)

[42] The parallel is just as striking for real numbers: I say the space-time points have the property of continuity, which can be detected using various set theoretic constructions; Shapiro says they exemplify the 'structure of Euclidean geometry' (Shapiro (1983*a*), 540), which is also exemplified by various set theoretic constructions.

[43] There has been some evolution in Resnik's thinking here. The account in Resnik (1975) suggests that we perceive the pattern itself; in Resnik (1982), we see the physical things and abstract the pattern. In Resnik (forthcoming *a*), this abstractionist epistemology is abandoned altogether in favour of a yet-to-be-developed 'postulational' view. Shapiro (1983*a*, p. 535) sticks with the abstractionist mode.

[44] Shapiro's account of our knowledge of infinite structures reads somewhat differently. See Shapiro (forthcoming).

Thus the structuralist's epistemology parallels the two-tiered account of the set theoretic realist. At the most elementary level, both theorists turn to perceptual knowledge—of sets or patterns—and after that, to theoretical knowledge, justified by its coherence and its consequences for lower-level theory.

So far, then, the structuralist and the set theoretic realist are in broad ontological and epistemological agreement: they meet the problem of multiple reductions of number theory with a move from numbers as objects to numbers as universals and the epistemological problem for traditional Platonism with a two-tiered epistemology of perceptual and theoretical justification. This only covers the natural numbers, but both advocate the same sort of exchange—objects for universals—for the reals, and presumably for other traditional mathematical objects that can be thought of as universals, multiply instantiated in the iterative hierarchy. Where the two part company, then, is in their view of the set theoretic universe itself. For the set theoretic realist, this 'structure' consists of real objects, the sets; these are the bedrock, the things that instantiate the various mathematical universals. For the structuralist, it is just one more structure, made up of featureless points in certain relations.

Though this surely sounds like a substantive disagreement, its true significance is difficult to assess. To see this, consider the structuralist's account of the interconnections between branches of mathematics. In order to explain, for example, how the study of the natural number structure can be advanced by study of the real number structure (in analytic number theory) or the iterative hierarchy structure (in reductions of number theory to set theory), the structuralist must speak of one structure being 'contained' or 'modelled' in another. For such purposes and others—e.g. for posing the question of whether or not $V = L$—the structuralist must speak of several structures at once and of the relations between them. This in turn requires an overarching 'structure theory'.

Of course, set theory can provide such a background theory; all structures can be taken to be sets (or proper classes of the least problematic kind), as can the functions and relations between them. This is the set theoretic realist's position. But the thoroughgoing structuralist would insist on a yet-to-be-described structure theory strong enough to encompass all structures, including the iterative hierarchy structure. The set theoretic partisan might wonder what such an all-encompassing structure theory would be like, and what could make it preferable to the more familiar theory of sets, but in fact, the prior question is: what would make these two theories different? Shapiro concludes that

> In a sense, the theories [set theory and a comprehensive structure theory] are notational variants of each other. (Shapiro (forthcoming))

If this is so, the purported difference between set theory as pattern and set theory as bedrock begins to elude us, along with that between structuralism and set theoretic realism.

Just for the record, I'd like to mention here two considerations that incline me to resist, at least for now, the characterization of set theoretic realism as a form of structuralism. The first is an epistemological disanalogy between arithmetic—a case for which even the set theoretic realist adopts a structuralist approach—and set theory—the case still open to debate. Structuralism for the natural numbers is so appealing partly because our understanding of arithmetic doesn't depend on which instantiation of the number structure we choose to study. For the purposes of simple perceptual access, as Resnik notes, a pattern of dots will do, as will a sequence of squares, stars, houses, coins, etc. The structuralist might say the same for set theory, that it matters not whether we begin from an array of dots, coins, or whatever, as long as they instantiate the initial stages of the iterative hierarchy pattern. For example, Mark Steiner, another thinker with strong structuralist tendencies,[45] writes:

> One imagines or looks at material bodies, and then diverts one's attention from their concrete spatial arrangement. . . . This is how one might become familiar with the standard model of ZF set theory—by abstracting from dots on a blackboard arranged in a certain way. Thus one arrives at an intuition of the structure of ZF sets. (Steiner (1975*a*), 134–5)

But I think it is not, in fact, the properties of such physical arrays that give us access to the simplest of set theoretic truths. Experience with any endless row might lead us to think that every number has

[45] See Steiner (1975*a*), 134. As noted above (ch. 3, sect. 1), Steiner's structuralism is only epistemic: numbers are objects, but the only things worth knowing about them are their relations to other numbers.

a successor, but it is experience with sets themselves that produces the intuitive belief that any two things can be collected into a set or that a set will have the same number of elements even after it has been rearranged. In other words, though any instantiation of the natural number structure can give us access to information about that structure, our information about the set theoretic hierarchy structure comes from our experience with one particular instantiation.[46] Thus one motivation for the move to structuralism in the case of number theory is undercut in the case of set theory.

My second source of concern about the assimilation of set theoretic realism to structuralism arises out of the simple question of what set theory is about. The set theoretic realist answers that set theory is the study of the iterative hierarchy with physical objects as ur-elements; the set theoretic monist takes physical objects themselves to be sets and eschews pure sets altogether. The trouble with these answers from the structuralist perspective is that some of the 'positions', in particular the ur-elements, have properties beyond those they have solely by virtue of their relations with other positions in the structure. The purely relational structure arising from the iterative hierarchy with ur-elements would make no distinction between the position occupied by this apple and the position occupied by this orange, between the position occupied by the set of the apple and the orange and the position occupied by the set of the apple and this baseball, distinctions the set theoretic realist will certainly want to preserve. Thus the structuralist bent on assimilating set theoretic realism by claiming that the iterative hierarchy with ur-elements is itself a purely relational structure will have to move to a larger, containing pattern, from whose point of view the baseball and the fruits are just positions with only relational properties.[47] My worry is how to square this with the naturalist's common-sense realism.[48]

But whatever the upshot of these inconclusive speculations about whether set theoretic realism should or shouldn't be considered a version of structuralism, my main goal here is to call attention to yet another point of agreement. Notice that in the pure iterative

[46] Parsons (forthcoming), § 9, makes a related point about the epistemological importance of recognizing non-relational features of sets.

[47] Resnik (personal communication) has suggested this move, and there are hints of it in Shapiro (forthcoming).

[48] See ch. 1, sect. 2, above.

hierarchy structure, the continuum hypothesis is either true or false, the projective sets either do or don't include a non-measurable set or an uncountable set without a perfect subset. Thus, only one of SC and V = L can be true there. So again, as in the previous section, the non-partisan conclusion is that structuralists, as well as compromise Platonists and Fieldian nominalists, will have to face the difficult problem of assessing the rationality of arguments for and against the various theoretical hypotheses that might answer these open questions.

There is a variation on structuralism according to which mathematics is the study not of structures but of possible structures. Rather than investigating sets (compromise Platonism) or the set theoretic pattern (structuralism), the modalist investigates what *would* be the case if there *were* a set theoretic hierarchy of the sort the Platonist describes.[49] '2 + 2 = 4' translates to 'if there were a natural number structure, 2 plus 2 would equal 4 in that structure'. This view has obvious if-thenist elements, and it suffers from many of the same difficulties.[50] Notice also that the modalist's actual world is purely physical; all mathematical things exist (if at all) in some other possible world. Even if the extreme monism of section 1 above is rejected, the pro-Platonist arguments of Quine and Putnam suggest that such a separation of the physical from the mathematical is not feasible. Thus the modalist, like Field, must find a way to defuse the indispensability arguments.[51]

Epistemologically, the modalist owes an account of modal knowledge that has not been forthcoming. One might think all that's needed is an explanation of the logical implication from, say, the Peano axioms to 2 + 2 = 4, but there is more; the modal translation will not work properly unless the Peano axioms are jointly possible.[52] In set theory, the corresponding requirement is that the iterative hierarchy be possible, and in this possible

[49] This sort of translation is suggested in Putnam (1967*a*), though he doesn't espouse modalism. Following Putnam's method, Hellman (1989) does.

[50] For a partial list of these, see ch. 1, sect. 4, above. For more, see Maddy (forthcoming *b*) or Resnik (1980), ch. 3.

[51] See Hellman (1989), ch. 3, and Field (1988), §§6–7, for an assessment of the modalist's prospects.

[52] If no such structure is even possible, 2 + 2 = 4 and 2 + 2 = 5 are both true, along with everything else. Again, the logic involved must be second-order if we are to speak of a unique natural number structure.

structure, the continuum hypothesis is either true or false, the projective sets either do or don't include a well-ordering of the reals, and so on. Thus, the modalist faces a question analogous to Field's—which of V = L or SC is conservative?—namely, which of these patterns is possible? I bring up modalism here primarily to point out that it is among the many positions, both Platonistic and non-Platonistic, that face not just the difficult question of whether or not a supercompact cardinal exists (here or in another possible world), but the prior and perhaps more difficult problem of how one might rationally answer such a question.

4. Summary

Realism about a given branch of inquiry is the contention that its subject matter exists objectively, that various efforts to reinterpret its claims should be resisted, and that most of its well-supported hypotheses are at least approximately true. I've endorsed common-sense realism about medium-sized physical objects on the grounds that the best explanation of why it seems to us that there is an objective world of such objects is that there is an objective world of such objects that is responsible for our beliefs. This explanation takes place, not within a priori philosophy, but within our scientific theory of the world and ourselves as cognizers; this is naturalism. I've also adopted scientific realism about the theoretical entities of natural science, because these unobservable things play a role in our best theory of the world. Similar reasoning cites the central role of classical mathematics in both the statement and the development of natural science as evidence for mathematical realism or Platonism. These are the pro-Platonist indispensability arguments of Quine and Putnam.

Quine/Putnam Platonism differs from the traditional variety over the purported a priority, certainty, and necessity of mathematical truth. As a complete theory of mathematical knowledge, it also differs from the practice of mathematics itself: it fails to account for unapplied mathematics and for the obviousness of elementary mathematics; it ignores the actual justificatory practices of mathematicians. Gödel's version of Platonism, by contrast, presents an appealing two-tiered account of justification within mathematics—intuitive and theoretical—but fails to support the scientific status of mathematics as a whole and rests its account of elementary knowledge on an unpersuasive notion of mathematical intuition. Nevertheless, Gödelian Platonism stands with Quine/Putnamism in opposition to the traditional variety.

I've proposed a compromise between these two modern versions of Platonism. From Quine/Putnamism, it takes the indispensability arguments as supports for the (approximate) truth of classical mathematics. From Gödel, it takes the two-tiered analysis of mathematical justification. But to provide a complete picture, compromise Platonism owes a replacement for Gödel's intuition; in deference to naturalism, this replacement must be scientifically feasible. The leading theme of this book has been the development and defence of set theoretic realism, a version of compromise Platonism designed to fill in this outline.

It has long been thought that Gödel's intuition, his epistemological bridge between the objects of mathematical knowledge and the mathematical knower, cannot be developed naturalistically. Benacerraf's classical statement of this worry (1973) depends on the then-popular causal theories of knowledge and reference, but I've argued that neither these nor a particularly robust notion of truth are essential to posing the problem. What matters is that the beliefs of mathematicians are reliable indicators of facts about mathematical things; this fact calls out for a naturalistic explanation. From this point, various forces—among them the conviction that mathematics is a legitimate science analogous to the physical sciences—lead to the conviction that at least part of this explanation must involve a perception-like connection between object known and knower. Add to this the traditional Platonist's characterization of mathematical entities as non-spatio-temporal and acausal, and it's easy to see why a naturalistic account is often considered impossible.

The set theoretic realist meets this problem by admitting sets of physical objects to the physical world, giving them spatio-temporal location where the physical stuff that makes up their members (and the members of their members, etc.) is located. These impure sets then prove appealing candidates for the subjects of perceptual numerical beliefs, and psychological and speculative neurological considerations give scientific support to the view that they are directly perceived. Thus part of the set theoretic realist's perception-like connection is just perception itself. An accompanying neurological phenomenon furnishes a rudimentary intuitive faculty whose products—intuitive beliefs—provide fallible but prima-facie justifications for the most elementary general assumptions of set theory.

This account depends essentially on the close relationship between numerical beliefs and beliefs about sets, which raises the familiar ontological question of whether numbers simply are sets. Part of the scientific support for set theory rests on the foundation it provides for number theory and analysis, and this foundational theory is standardly expressed by identifying the natural and real numbers with certain sets. But, as Benacerraf has pointed out, this identification is ultimately unsatisfying because it can be done with equal ease in several different ways; this is the problem of multiple reductions. If there is nothing to decide between the von Neumann and the Zermelo ordinals when identifying the natural numbers with sets, how can either sequence of sets claim to actually be the numbers? The set theoretic realist's answer, implicit in the account of set perception, is that neither sequence is the numbers, that numbers are properties of sets which either sequence is equally well equipped to measure. The same line of response works for the real numbers when they are understood as detectors for the property of continuity.

If the first tier of Gödel's epistemological theory can be ascribed to the set theoretic realist's perception and intuition, there remains the problem of describing and accounting for the rationality of reasoning at the theoretical level. In set theory, despite traces of the traditional Platonistic view that axioms are obvious or self-evident, theoretical defences for axiom candidates can be found even in Zermelo's first axiomatization, and they figure prominently in the search for new hypotheses that will decide natural analytic and set theoretic questions left open by the currently accepted axioms of ZFC. The problem of assessing the rationality of various non-demonstrative arguments for and against new set theoretic hypotheses becomes more acute as set theorists devise alternative, conflicting, theories. The first step in helping adjudicate such disputes is a descriptive catalogue of the evidence offered by each side. A modest contribution to that project is all that has been attempted here. The next step, the evaluation of this evidence, is a daunting undertaking, but I've argued that the set theoretic realist faces this challenge in the distinguished company of thinkers representing a wide range of competing mathematical philosophies, structuralism, modalism, and a version of nominalism among them.

Finally, for the benefit of those with physicalistic leanings, I've sketched set theoretic monism, a minor variation on set theoretic realism. For the monist, all sets have physical grounding and spatio-temporal location, and all physical objects are sets. These manœvres produce a radical 'one-worldism'—a reality at once mathematical and physical—that should appeal to philosophers of this stripe.

In sum, then, I certainly do not claim to have shown that my version of Platonism raises no difficult philosophical problems. At best, *at best*, I have shown how to replace the two prominent Benacerraf-style objections to traditional Platonism with a new open question about the justification of theoretical hypotheses in set theory. But whatever the complexities of this new problem, I think this trade amounts to progress. In the defence of mathematical realism, the new problem enjoys a clear advantage over its predecessors: nothing on its face is likely to inspire one of those nagging a priori arguments against the very possibility of Platonism. On the contrary, the questions it raises—questions of rationality—are standard in the philosophy of all sciences, and there is no obvious reason why they should be any less tractable in mathematics than they are in physics or physiology.

But there is more to be said for this new problem than that it may lighten the perceived burden on the defender of Platonism. I attach considerable importance to the fact that it arises also for adherents of alternative philosophical positions; this suggests that it taps into a fundamental issue insensitive to minor variations in philosophical fashion. And beyond this, there is the alluring possibility that philosophical progress on questions of mathematical rationality could make a real contribution to mathematics itself, especially to the current search for new axioms. Thus, once again, I recommend pursuit of this new problem even to philosophers blissfully uninvolved in the debate over Platonism.

Mathematicians often think of themselves as scientists, exploring the intricacies of mathematical reality; and, for good reason, they are especially inclined towards such views in the absence of philosophers. I have tried to show that, contrary to popular philosophical opinion, something close to the mathematician's natural attitude is defensible. Theories of mathematical knowledge tend either to trivialize it as conventional or purely formal or even false, or to glamorize it as perfect, a priori, and certain, but set theoretic realism aims to treat it as no more nor less than the science it is, and to be fair, all at once, to the mathematician who produces the knowledge, the scientist who uses it, and the cognitive scientist who must explain it. I propose it, then, as another step—after Gödel, Quine, and Putnam—on the long road towards mathematics naturalized.

REFERENCES

ACHINSTEIN, P. (1965), 'The problem of theoretical terms', repr. in Brody (ed.) (1970), 234–50.

ACKERMANN, W. (1956), 'Zur Axiomatik der Mengenlehre', *Mathematische Annalen*, 131, pp. 336–45.

ACZEL, P. (1988), *Non-Well-Founded Sets*, Center for the Study of Language and Information, Lecture Notes, no. 14.

ADDISON, J. W. (1958), 'Separation principles in the hierarchies of classical and effective descriptive set theory', *Fundamenta Mathematicae*, 46, pp. 123–35.

—— (1959), 'Some consequences of the axiom of constructibility', *Fundamenta Mathematicae*, 46, pp. 337–57.

—— and MOSCHOVAKIS, Y. N. (1968), 'Some consequences of the axiom of definable determinateness', *Proceedings of the National Academy of Sciences (U. S. A.)*, 59, pp. 708–12.

ALEXANDROFF, P. (1916) 'Sur la puissance des ensembles mesurables B', *Comptes rendus de l'Académie des Sciences de Paris*, 162, pp. 232–5.

ANDERSON, C. A. (1987), 'Review of Bealer's *Quality and Concept*', *Journal of Philosophical Logic*, 16, pp. 115–64.

ARISTOTLE (1952), *The Works of Aristotle Translated into English*, 12 vols., ed. W. D. Ross (Oxford: Oxford University Press).

—— *Categories*, in his (1952).

—— *Metaphysics*, in his (1952).

—— *Physics*, in his (1952).

ARMSTRONG, D. (1961), *Perception and the Physical World* (London: Routledge and Kegan Paul).

—— (1973), *Belief, Truth and Knowledge* (Cambridge: Cambridge University Press).

—— (1977), 'Naturalism, materialism and first philosophy', repr. in his (1981), 149–65.

—— (1978), *Universals and Scientific Realism* (Cambridge: Cambridge University Press).

—— (1980), 'Against "ostrich nominalism".', *Pacific Philosophical Quarterly*, 16, pp. 440–9.

—— (1981), *The Nature of Mind* (Ithaca, NY: Cornell University Press).

AYER, A. J. (1946), *Language, Truth, and Logic*, 2nd edn. (New York: Dover, 1952).

AYERS, M. R. (1981), 'Locke versus Aristotle on natural kinds', *Journal of Philosophy*, 78, pp. 247–72.

BAIRE, R. (1899). 'Sur les fonctions de variables réelles', *Annali di matematica pura ed applicata*, 3, pp. 1–122.

—— BOREL, E., HADAMARD, J., and LEBESGUE, H. (1905), 'Five letters on set theory', repr. in Moore (1982), 311–20.

BARWISE, J. (ed.) (1977), *The Handbook of Mathematical Logic* (Amsterdam: North Holland).

BEALER, G. (1982), *Quality and Concept* (Oxford: Oxford University Press).

BENACERRAF, P. (1965), 'What numbers could not be', repr. in Benacerraf and Putnam (eds.) (1983), 272–94.

—— (1973), 'Mathematical truth', repr. in Benacerraf and Putnam (eds.) (1983), 403–20.

—— (1985), 'Comments on Maddy and Tymoczko', in Kitcher (ed.) (1985), 476–85.

—— and PUTNAM, H. (eds.) (1983), *Philosophy of Mathematics*, 2nd edn. (Cambridge: Cambridge University Press).

BERKELEY, G. (1710), *The Principles of Human Knowledge*, in his (1957).

—— (1713), *Three Dialogues between Hylas and Philonous*, in his (1957).

—— (1734), *The Analyst*, in his (1957).

—— (1957), *The Works of George Berkeley, Bishop of Cloyne*, 9 vols., ed. A. Luce and T. Jessop (London: Thomas Nelson and Sons).

BERNAYS, P. (1935), 'On platonism in mathematics', repr. in Benacerraf and Putnam (eds.) (1983), 258–71.

—— (1937), 'A system of axiomatic set theory, I', *Journal of Symbolic Logic*, 2, pp. 65–77.

BLACKWELL, D. (1967), 'Infinite games and analytic sets', *Proceedings of the National Academy of Sciences (U.S.A.)*, 58, pp. 1836–7.

BONEVAC, D. A. (1982), *Reduction in the Abstract Sciences* (Indianapolis, Ind.: Hackett).

BONJOUR, L. (1980), 'Externalist theories of empirical knowledge', *Midwest Studies in Philosophy*, 5 (Minneapolis: University of Minnesota Press), 53–73.

BOOLOS, G. (1971), 'The iterative conception of set', repr. in Benacerraf and Putnam (eds.) (1983), 486–502.

BOREL, E. (1898), *Leçons sur la théorie des fonctions* (Paris: Gauthier-Villars).

BOWER, T. G. R. (1966), 'The visual world of infants', *Scientific American*, 215, no. 6, pp. 80–92.

—— (1982), *Development in Infancy*, 2nd edn. (San Francisco: W. H. Freeman and Company).

BOYER, C. B. (1949), *The History of the Calculus and its Conceptual Development*, (New York: Dover, 1959). (Original title: *The Concepts of the Calculus.*)

BRIDGMAN, P. W. (1927), *The Logic of Modern Physics* (New York: Macmillan).

BRODY, B. (ed.) (1970), *Readings in the Philosophy of Science* (Englewood Cliffs, NJ: Prentice Hall).

BROUWER, L. E. J. (1913), 'Intuitionism and formalism', repr. in Benacerraf and Putnam (eds.) (1983), 77–89.

—— (1949), 'Consciousness, philosophy, and mathematics', repr. in Benacerraf and Putnam (eds.) (1983), 90–6.

BRUNER, J. (1957), 'On perceptual readiness', repr. in R. Harper *et al.* (eds.), *The Cognitive Processes* (Englewood Cliffs, NJ: Prentice Hall, 1964), 225–56.

BURALI-FORTI, C. (1897), 'A question on transfinite numbers', repr. in van Heijenoort (ed.) (1967), 104–12.

BURGESS, J. P. (1983). 'Why I am not a nominalist', *Notre Dame Journal of Formal Logic*, 24, pp. 93–105.

—— (1984), 'Synthetic mechanics', *Journal of Philosophical Logic*, 13, pp. 379–95.

—— (forthcoming *a*), 'Synthetic physics and nominalist realism', to appear in C. W. Savage and P. Erlich (eds.), *The Nature and Function of Measurement.*

—— (forthcoming *b*), 'Epistemology and nominalism', to appear in A. Irvine (ed.), *Physicalism in Mathematics.*

CANTOR, G. (1872), 'Über die Ausdehnung eines Satzes aus der Theorie der trigonometrischen Reihen', *Mathematische Annalen*, 5, pp. 123–32.

—— (1878), 'Ein Beitrag zur Mannigfaltigkeitslehre', *Journal für die reine und angewandte Mathematik*, 84, pp. 242–58.

—— (1883), *Grundlagen einer allgemeinen Mannigfaltigkeitslehre* (Leipzig: B. G. Teubner).

—— (1891), 'Über eine elementare Frage der Mannigfaltigkeitslehre', *Jahresbericht der Deutschen Mathematiker-Vereinigung*, 1, pp. 75–8.

—— (1895/7), *Contributions to the Founding of the Theory of Transfinite Numbers*, ed. P. E. B. Jourdain (Chicago: Open Court, 1915).

—— (1899), 'Letter to Dedekind', repr. in van Heijenoort (ed.) (1967), 113–17.

CARNAP, R. (1934), *The Unity of Science* (London: Kegan Paul, Trench, Trubner and Company).

—— (1936/7), 'Testability and meaning', *Philosophy of Science*, 3, pp. 428–68, and 4, pp. 1–40.

—— (1937), *Logical Syntax of Language* (London: Routledge and Kegan Paul).

—— (1950), 'Empiricism, semantics, and ontology', repr. in Benacerraf and Putnam (eds.) (1983), 241–57.

CASULLO, A. (forthcoming) 'Causality, reliabilism, and mathematical knowledge', to appear.

CHIHARA, C. (1973), *Ontology and the Vicious-Circle Principle* (Ithaca, NY: Cornell University Press).

—— (1982), 'A Gödelian thesis regarding mathematical objects: Do they exist? And can we perceive them?', *Philosophical Review*, 91, pp. 211–27.

CHISHOLM, R. (1977), *Theory of Knowledge*, 2nd edn. (Englewood Cliffs, NJ: Prentice Hall).

COHEN, P. J. (1966), *Set Theory and the Continuum Hypothesis* (New York: W. A. Benjamin).

DAUBEN, J. W. (1979), *Georg Cantor* (Cambridge, Mass.: Harvard University Press).

DAVIS, M. (1964), 'Infinite games of perfect information', *Annals of Mathematics Studies*, 52, pp. 85–101.

DAVIS, P. J., and HERSH, R. (1981), *The Mathematical Experience* (Boston: Birkhauser).

DEDEKIND, R. (1872), 'Continuity and irrational numbers', in *Essays on the Theory of Numbers* (La Salle, Ill.: Open Court, 1901), pp. 1–27.

DENNETT, D. C. (1978), *Brainstorms* (Bradford Books).

DESCARTES, R. (1641), *Meditations on First Philosophy*, 2nd edn., in his (1967).

—— (1967), *Philosophical Works of Descartes*, 2 vols., ed. E. S. Haldane and G. R. T. Ross (Cambridge: Cambridge University Press).

DETLEFSEN, M. (1986), *Hilbert's Program* (Dordrecht: Reidel).

DEVITT, M. (1980), '"Ostrich nominalism" or "mirage realism"?', *Pacific Philosophical Quarterly*, 61, pp. 433–9.

—— (1981), *Designation* (New York: Columbia University Press).

—— (1984), *Realism and Truth* (Princeton, NJ: Princeton University Press).

DEVLIN, K. (1977), *The Axiom of Constructibility* (Berlin: Springer–Verlag).

—— (1984), *Constructibility* (Berlin: Springer-Verlag).

DRAKE, F. (1974), *Set Theory: An Introduction to Large Cardinals* (Amsterdam: North Holland).

DUMMETT, M. (1975), 'The philosophical basis of intuitionist logic', repr. in his (1978), ch. 14, and in Benacerraf and Putnam (eds.) (1983), 97–129.

—— (1977), *Elements of Intuitionism* (Oxford: Oxford University Press).

—— (1978), *Truth and Other Enigmas* (Cambridge, Mass.: Harvard University Press).

Eklof, P. C., and Mekler, A. H. (*forthcoming*), *Almost Free Modules: Set-Theoretic Methods*, *forthcoming* from North Holland Publishers, Mathematical Library Series.

Ellis, B. (1966), *Basic Concepts of Measurement* (Cambridge: Cambridge University Press).

Enderton, H. (1972), *A Mathematical Introduction to Logic* (New York: Academic Press).

—— (1977), *Elements of Set Theory* (New York: Academic Press).

Feferman, S. (1984*a*), 'Toward useful type-free theories, I', *Journal of Symbolic Logic*, 49, pp. 75–111.

—— (1984*b*), 'Kurt Gödel: Conviction and caution', repr. in S. G. Shanker (ed.), *Gödel's Theorem in Focus* (London: Croom Helm, 1988), 96–114.

—— (1988), 'Hilbert's program relativized: Proof-theoretical and foundational reductions', *Journal of Symbolic Logic*, 53, pp. 364–84.

Field, H. (1972), 'Tarski's theory of truth', *Journal of Philosophy*, 69, pp. 347–75.

—— (1980), *Science without Numbers* (Princeton, NJ: Princeton University Press).

—— (1982), 'Realism and anti-realism about mathematics', repr. in his (1989), 53–78.

—— (1984), 'Is mathematical knowledge just logical knowledge?', repr. in his (1989), 79–124.

—— (1985), 'On conservativeness and incompleteness', repr. in his (1989), 125–46.

—— (1986), 'The deflationary conception of truth', in G. MacDonald and C. Wright (eds.), *Fact, Science and Value* (Oxford: Basil Blackwell), 55–117.

—— (1988), 'Realism, mathematics, and modality', repr. in his (1989), 227–81.

—— (1989), *Realism, Mathematics, and Modality* (Oxford: Basil Blackwell).

—— (forthcoming), 'Physicalism'.

Fodor, J. A. (1975), *The Language of Thought* (New York: Thomas Y. Crowell).

Foreman, M. (1986), 'Potent axioms', *Transactions of the American Mathematical Society*, 294, pp. 1–28.

Fraenkel, A. A. (1922), 'Zu den Grundlagen der Cantor-Zermeloschen Mengenlehre', *Mathematische Annalen*, 86, pp. 230–7.

—— Bar-Hillel, Y., and Levy, A. (1973), *Foundations of Set Theory*, 2nd rev. edn. (Amsterdam: North Holland).

Frege, G. (1884), *The Foundations of Arithmetic*, 2nd rev. edn. (Evanston, Ill: Northwestern University Press, 1968).

—— (1892*a*), 'On concept and object', in his (1970), 42–55.

—— (1892*b*), 'On sense and reference', in his (1970), 56–78.
—— (1903), *Grundgesetze der Arithmetik*, vol. ii. Relevant sections are reprinted in his (1970), 182–233.
—— (1970), *Translations from the Philosophical Writings of Gottlob Frege*, ed. P. Geach and M. Black (Oxford: Basil Blackwell).
—— (1979), *Posthumous Writings*, ed. H. Hermes, F. Kambartel, and F. Kaulbach (Chicago: University of Chicago Press).
FRIEDMAN, H. (1971), 'Higher set theory and mathematical practice', *Annals of Mathematical Logic*, 2, pp. 325–57.
GALE, D., and STEWART, F. M. (1953), 'Infinite games with perfect information', *Annals of Mathematics Studies*, 28, pp. 245–66.
GELMAN, R. (1977), 'How young children reason about small numbers', in N. Castellan, D. Pisoni, and G. Potts (eds.), *Cognitive Theory*, ii (Hillsdale, NJ: Lawrence Erlbaum Associates), 219–38.
GETTIER, E. (1963), 'Is justified true belief knowledge?', *Analysis*, 23, pp. 121–3.
GIBSON, E. (1969), *Principles of Perceptual Learning and Development* (New York: Appleton-Century-Crofts).
GIBSON, J. J. (1950), *The Perception of the Visual World* (Boston: Houghton Mifflin).
GÖDEL, K. (1930), 'The completeness of the axioms of the functional calculus of logic', repr. in van Heijenoort (ed.) (1967), 582–91.
—— (1931), 'On formally undecidable propositions of *Principia Mathematica* and related systems, I', repr. in van Heijenoort (ed.) (1967), 596–616.
—— (1938), 'The consistency of the axiom of choice and of the generalized continuum hypothesis', *Proceedings of the National Academy of Sciences (U.S.A.)*, 24, pp. 556–7.
—— (1940), *The Consistency of the Continuum Hypothesis* (Princeton: Princeton University Press).
—— (1944), 'Russell's mathematical logic', repr. in Benacerraf and Putnam (eds.) (1983), 447–69.
—— (1946), 'Remarks before the Princeton Bicentennial Conference on problems in mathematics', in M. Davis (ed.), *The Undecidable* (New York: Raven Press, 1965), 84–8.
—— (1947/64), 'What is Cantor's continuum problem?', repr. in Benacerraf and Putnam (eds.) (1983), 470–85.
GOLDMAN, A. (1967), 'A causal theory of knowing', *Journal of Philosophy*, 64, pp. 357–72.
—— (1975), 'Innate knowledge', in S. Stich (ed.), *Innate Ideas* (Berkeley: University of California Press), 111–20.
—— (1976), 'Discrimination and perceptual knowledge', *Journal of Philosophy*, 73, pp. 771–91.
—— (1977), 'Perceptual objects', *Synthese*, 35, pp. 257–84.

GOLDMAN, A. (1979), 'What is justified belief?', in G. Pappas (ed.), *Justification and Knowledge* (Amsterdam: Reidel), 1–23.

—— (1980), 'The internalist conception of justification', *Midwest Studies in Philosophy*, 5 (Minneapolis: University of Minnesota Press), 27–51.

GOTTLIEB, D. (1980), *Ontological Economy* (Oxford: Oxford University Press).

GREGORY, R. L. (1970), *The Intelligent Eye* (New York: McGraw-Hill).

—— (1972), *Eye and Brain*, 2nd edn. (New York: McGraw-Hill).

GRICE, P. (1961), 'The causal theory of perception', repr. in R. Swartz (ed.), *Perceiving, Sensing and Knowing* (Berkeley: University of California Press), 438–72.

GROVER, D., CAMP, J., and BELNAP, N. (1975), 'A prosentential theory of truth', *Philosophical Studies*, 27, pp. 73–125.

HALE, B. (1987). *Abstract Objects* (Oxford: Basil Blackwell).

HALLETT, M. (1984), *Cantorian Set Theory and Limitation of Size* (Oxford: Oxford University Press).

HAMBOURGER, R. (1977), 'A difficulty with the Frege–Russell definition of number', *Journal of Philosophy*, 74, pp. 409–14.

HARMAN, G. (1973), *Thought* (Princeton: Princeton University Press).

HARRINGTON, L. A., MORLEY, M. D., SCEDROV, A., and SIMPSON, S. G. (eds.) (1985), *Harvey Friedman's Research on the Foundations of Mathematics* (Amsterdam: North Holland).

HART, W. H. (1977), 'Review of Steiner's *Mathematical Knowledge*', *Journal of Philosophy*, 74, pp. 118–29.

HAUSDORFF, F. (1919), 'Über halbstetige Funktionen und deren Verallgemeinerung, *Mathematische Zeitschrift*, 5, pp. 292–309.

HEBB, D. O. (1949), *The Organization of Behavior* (New York: John Wiley and Sons).

—— (1980), *Essay on Mind* (Hillsdale, NJ: Lawrence Erlbaum Associates).

HELLMAN, G. (1989), *Mathematics Without Numbers* (Oxford: Oxford University Press).

HEMPEL, C. G. (1945), 'On the nature of mathematical truth', repr. in Benacerraf and Putnam (eds.) (1983), 377–93.

—— (1954), 'A logical appraisal of operationalism', repr. in his (1965), 123–33.

—— (1965), *Aspects of Scientific Explanation* (New York: The Free Press).

HENKIN, L. (1949), 'The completeness of the first-order functional calculus', *Journal of Symbolic Logic*, 14, pp. 159–66.

HEYTING, A. (1931), 'The intuitionist foundations of mathematics', repr. in Benacerraf and Putnam (eds.) (1983), 52–61.

—— (1966), *Intuitionism: An Introduction*, 2nd rev. edn. (Amsterdam: North Holland).

HILBERT, D. (1899), *Foundations of Geometry* (La Salle, Ill.: Open Court, 1971).
—— (1926), 'On the infinite', repr. in Benacerraf and Putnam (eds.) (1983), 183–201, and in van Heijenoort (ed.) (1967), 367–92.
—— (1928), 'The foundations of mathematics', repr. in van Heijenoort (ed.) (1967), 464–79.
HODES, H. (1984), 'Logicism and the ontological commitments of arithmetic', *Journal of Philosophy*, 81, pp. 123–49.
HUME, D. (1739), *A Treatise of Human Nature*, vol. i, in his (1886).
—— (1886), *Philosophical Works*, 4 vols., ed. T. H. Green and T. H. Grose (London).
JECH, T. (1978), *Set Theory* (New York: Academic Press).
JUBIEN, M. (1977), 'Ontology and mathematical truth', *Nous*, 11, pp. 133–50.
KATZ, J. J. (1981), *Language and Other Abstract Objects* (Totowa, NJ: Rowman and Littlefield).
KAUFMAN, E. L., LORD, M. W., REESE, T. W., and VOLKMANN, J. (1949), 'The discrimination of visual number', *American Journal of Psychology*, 62, pp. 498–525.
KELLEY, J. L. (1955), *General Topology* (Princeton, NJ: van Nostrand).
KIM, J. (1977), 'Perception and reference without causality', *Journal of Philosophy*, 74, pp. 606–20.
—— (1981), 'The role of perception in a priori knowledge', *Philosophical Studies*, 40, pp. 339–54.
KITCHER, P. (1978), 'The plight of the Platonist', *Nous*, 12, pp. 119–36.
—— (1983), *The Nature of Mathematical Knowledge* (New York: Oxford University Press).
—— (ed.) (1985), *PSA 1984*, ii (East Lansing: Philosophy of Science Association).
KLINE, M. (1972), *Mathematical Thought from Ancient to Modern Times* (New York: Oxford University Press).
KORNER, S. (1960), *The Philosophy of Mathematics* (London: Hutchinson University Library).
KRIPKE, S. (1972), 'Naming and necessity', in D. Davidson and G. Harman (eds.), *Semantics of Natural Language* (Dordrecht: Reidel), 253–355, 763–9.
—— (1975), 'Outline of a theory of truth', *Journal of Philosophy*, 72, pp. 690–716.
—— (1982), *Wittgenstein on Rules and Private Language* (Cambridge, Mass.: Harvard University Press).
KURATOWSKI, K. (1966), *Topology*, i (New York: Academic Press).
LEAR, J. (1977), 'Sets and semantics', *Journal of Philosophy*, 74, pp. 86–102.

LEBESGUE, H. (1902), 'Intégrale, longueur, aire', *Annali di matematica pura ed applicata*, 7, pp. 231–359.

—— (1905), 'Sur les fonctions représentables analytiquement, *Journal de mathématiques pures et appliquées*, 60, pp. 139–216.

LEEDS, S. (1978), 'Theories of reference and truth', *Erkenntnis*, 13, pp. 111–29.

LETTVIN, J. Y., MATURANA, H. R., MCCULLOCH, W. S., and PITTS, W. H. (1959), 'What the frog's eye tells the frog's brain', repr. in W. S. McCulloch, *Embodiments of Mind* (Cambridge, Mass.: MIT Press, 1965), 230–55.

LEVY, A., and SOLOVAY, R. M. (1967), 'Measurable cardinals and the continuum hypothesis', *Israel Journal of Mathematics*, 5, pp. 234–48.

LEWIS, D. (1983), 'New work for a theory of universals', *Australian Journal of Philosophy*, 61, pp. 343–77.

—— (1984), 'Putnam's paradox', *Australian Journal of Philosophy*, 62, pp. 221–36.

—— (1986), *On the Plurality of Worlds* (Oxford: Basil Blackwell).

LOCKE, J. (1690), *An Essay Concerning Human Understanding* (New York: Dover, 1959).

LUCE, L. (1988), 'Frege on cardinality', *Philosophy and Phenomenological Research*, 48, pp. 415–34.

LUZIN, N. (1917), 'Sur la classification de M. Baire', *Comptes rendus de l'Académie des Sciences de Paris*, 164, pp. 91–4.

—— (1925), 'Sur les ensembles projectifs de M. Henri Lebesgue', *Comptes rendus de l'Académie des Sciences de Paris*, 180, pp. 1572–4.

—— (1927), 'Sur les ensembles analytiques', *Fundamenta Mathematicae*, 10, pp. 1–95.

MACHAMER, P. (1970), 'Recent work on perception', *American Philosophical Quarterly*, 7, pp. 1–22.

MADDY, P. (1980), 'Perception and mathematical intuition', *Philosophical Review*, 89, pp. 163–96.

—— (1981), 'Sets and numbers', *Nous*, 15, pp. 494–511.

—— (1983), 'Proper classes', *Journal of Symbolic Logic*, 48, pp. 113–39.

—— (1984*a*), 'Mathematical epistemology: what is the question?', *Monist*, 67, pp. 46–55.

—— (1984*b*), 'How the causal theorist follows a rule', *Midwest Studies in Philosophy*, 9 (Minneapolis: University of Minnesota Press), 457–77.

—— (1984*c*), 'Informal notes on proper classes', unpublished notes.

—— (1986), 'Mathematical alchemy', *British Journal for the Philosophy of Science*, 37, pp. 279–314.

—— (1988*a*), 'Believing the axioms', *Journal of Symbolic Logic*, 53, pp. 481–511, 736–64.

—— (1988*b*), 'Mathematical realism', *Midwest Studies in Philosophy*, 12 (Minneapolis: University of Minnesota Press), 275–85.

—— (forthcoming *a*), 'Physicalistic Platonism', to appear in A. Irvine (ed.), *Physicalism in Mathematics*.

—— (forthcoming *b*), 'The roots of contemporary Platonism', to appear in the *Journal of Symbolic Logic*.

MALAMENT, D. (1982), 'Review of Field's *Science Without Numbers*', *Journal of Philosophy*, 79, pp. 523–34.

MARTIN, D. A. (1968), 'The axiom of determinateness and reduction principles in the analytical hierarchy', *Bulletin of the American Mathematical Society*, 74, pp. 687–9.

—— (1970), 'Measurable cardinals and analytic games', *Fundamenta Mathematicae*, 66, pp. 287–91.

—— (1975), 'Borel determinacy', *Annals of Mathematics*, 102, pp. 363–71.

—— (1976), 'Hilbert's first problem: The continuum hypothesis', *Proceedings of Symposia in Pure Mathematics*, 28, (Providence, RI: American Mathematical Society), 81–92.

—— (1977), 'Descriptive set theory: projective sets', in Barwise (ed.) (1977), 783–815.

—— (1980), 'Infinite games', *Proceedings of the International Congress of Mathematicians (Helsinki, 1978)*, pp. 269–73.

—— (1985), 'A purely inductive proof of Borel determinacy', *Proceedings of Symposia in Pure Mathematics*, 42 (Providence, RI: American Mathematical Society), 303–8.

—— 'Projective sets and cardinal numbers', unpublished photocopy.

—— 'Sets versus classes', unpublished photocopy.

—— and SOLOVAY, R. M. (1970), 'Internal Cohen extensions', *Annals of Mathematical Logic*, 2, pp. 143–78.

—— and STEEL, J. (1988), 'Projective determinacy', *Proceedings of the National Academy of Sciences (U.S.A.)*, 85, pp. 6582–6.

—— —— (1989), 'A proof of projective determinacy', *Journal of the American Mathematical Society*, 2, pp. 71–125.

MAXWELL, G. (1962), 'The ontological status of theoretical entities', repr. in Brody (ed.) (1970), 224–33.

MENZEL, C. (1988), 'Frege numbers and the relativity argument', *Canadian Journal of Philosophy*, 18, pp. 87–98.

MERRILL, G. H. (1980), 'The model-theoretic argument against realism', *Philosophy of Science*, 47, pp. 69–81.

MILL, J. S. (1843), *A System of Logic*, in his (1963/88), vols. vii and viii.

—— (1865), *An Examination of Sir William Hamilton's Philosophy*, in his (1963/88), vol. ix.

—— (1963/88), *The Collected Works of John Stuart Mill*, 29 vols., ed. J. M. Robson and J. Stillinger (Toronto: University of Toronto Press).

MIRIMANOFF, D. (1917*a*), 'Les Antinomies de Russell et de Burali-Forti et le problème fondamental de la théorie des ensembles', *L'Enseignement mathématique*, 19, pp. 37–52.

MIRIMANOFF, D. (1917*b*), 'Remarques sur la théorie des ensembles et les antinomies Cantoriennes, I', *L'Enseignement mathématique*, 19, pp. 209–17.

MONNA, A. F. (1972), 'The concept of function in the 19th and 20th centuries', *Archive for History of Exact Sciences*, 9, pp. 57–84.

MOORE, G. H. (1982), *Zermelo's Axiom of Choice* (New York: Springer-Verlag).

—— (forthcoming), 'Introductory note to 1947 and 1964', *The Collected Works of Kurt Gödel*, vol. ii, forthcoming from Oxford University Press.

MORSE, A. (1965), *A Theory of Sets* (New York: Academic Press).

MOSCHOVAKIS, Y. N. (1970), 'Determinacy and prewellorderings of the continuum', in Y. Bar-Hillel (ed.), *Mathematical Logic and Foundations of Set Theory* (Amsterdam: North Holland), 24–62.

—— (1980), *Descriptive Set Theory* (Amsterdam: North Holland).

MYCIELSKI, J., and STEINHAUS, H. (1962), 'A mathematical axiom contradicting the axiom of choice', *Bulletin de l'Académie Polonaise des Sciences*, 10, pp. 1–3.

—— and ŚWIERCZKOWSKI, S. (1964), 'On the Lebesgue measurability and the axiom of determinateness', *Fundamenta Mathematicae*, 54, pp. 67–71.

NEISSER, U. (1976), *Cognition and Reality* (San Francisco: W. H. Freeman and Company).

NOVIKOV, P. (1935), 'Sur la séparabilité des ensembles projectifs de seconde classe', *Fundamenta Mathematicae*, 25, pp. 459–66.

NYIKOS, P. (forthcoming), 'Testimony on large cardinals and set-theoretic consistency results'.

PARSONS, C. (1965), 'Frege's theory of number', repr. in his (1983*a*), 150–75.

—— (1974*a*), 'Sets and classes', repr. in his (1983*a*), 209–20.

—— (1974*b*), 'The liar paradox', repr. in his (1983*a*), 221–67.

—— (1977), 'What is the iterative conception of set?', repr. in his (1983*a*), 268–97, and in Benacerraf and Putnam (eds.) (1983), 503–29.

—— (1979/80), 'Mathematical intuition', *Proceedings of the Aristotelian Society*, 80, pp. 145–68.

—— (1983*a*), *Mathematics in Philosophy* (Ithaca, NY: Cornell University Press).

—— (1983*b*), 'Quine on the philosophy of mathematics', in his (1983*a*), 176–205.

—— (forthcoming), 'The structuralist view of mathematical objects', to appear in *Synthese*.

PHILLIPS, J. L. (1975), *The Origins of Intellect: Piaget's Theory*, 2nd edn. (San Francisco: W. H. Freeman and Company).

PIAGET, J. (1937), *The Construction of Reality in the Child* (New York: Basic Books, 1954).

—— and INHELDER, B. (1948), *The Child's Conception of Space* (New York: W. W. Norton, 1967).
—— and SZEMIŃSKA, A. (1941), *The Child's Conception of Number* (New York: Humanities Press, 1952).
PITCHER, G. (1971), *A Theory of Perception* (Princeton, NJ: Princeton University Press).
PLATO (1871), *The Dialogues of Plato*, 5 vols., ed. B. Jowett (Oxford: Oxford University Press).
—— *Phaedo*, in his (1871).
—— *Phaedrus*, in his (1871).
—— *Republic*, in his (1871).
—— *Theaetetus*, in his (1871).
—— *Timaeus*, in his (1871).
PUTNAM, H. (1962), 'What theories are not', repr. in his (1979), 215–27.
—— (1967*a*), 'Mathematics without foundations', repr. in his (1979), 43–59, and in Benacerraf and Putnam (eds.) (1983), 295–311.
—— (1967*b*), 'The thesis that mathematics is logic', repr. in his (1979), 12–42.
—— (1968), 'The logic of quantum mechanics', repr. in his (1979), 174–97.
—— (1970), 'On properties', repr. in his (1979), 305–22.
—— (1971), 'Philosophy of logic', repr. in his (1979), 323–57.
—— (1975*a*), *Mind, Language and Matter* (Philosophical Papers, 2) (Cambridge: Cambridge University Press).
—— (1975*b*), 'What is mathematical truth?', repr. in his (1979), 60–78.
—— (1977), 'Realism and reason', repr. in his (1978), 123–38.
—— (1978), *Meaning and the Moral Sciences* (Boston: Routledge and Kegan Paul).
—— (1979), *Mathematics, Matter and Method* (Philosophical Papers, 1, 2nd edn.) (Cambridge: Cambridge University Press).
—— (1980), 'Models and reality', repr. in Benacerraf and Putnam (eds.) (1983), 421–44.
QUINE, W. V. O. (1936), 'Truth by convention', repr. in Benacerraf and Putnam (eds.) (1983), 329–54.
—— (1948), 'On what there is', repr. in his (1980*a*), 1–19.
—— (1951), 'Two dogmas of empiricism', repr. in his (1980*a*), 20–46.
—— (1954), 'Carnap and logical truth', repr. in Benacerraf and Putnam (eds). (1983), 355–76.
—— (1969*a*), *Set Theory and its Logic*, rev. edn. (Cambridge, Mass.: Harvard University Press).
—— (1969*b*), 'Epistemology naturalized', in his (1969*c*), 69–90.
—— (1969*c*), *Ontological Relativity* (New York: Columbia University Press).
—— (1969*d*), 'Natural kinds', in his (1969*c*), 114–38.

Quine, W. V. O. (1980*a*), *From a Logical Point of View*, 2nd edn., rev. (Cambridge, Mass.: Harvard University Press).

—— (1980*b*), 'Soft impeachment disowned', *Pacific Philosophical Quarterly*, 61, pp. 450–1.

—— (1984), 'Review of Parsons's *Mathematics in Philosophy*', *Journal of Philosophy*, 81, pp. 783–94.

Reinhardt, W. N. (1974), 'Set existence principles of Shoenfield, Ackermann, and Powell', *Fundamenta Mathematicae*, 84, pp. 5–34.

Resnik, M. (1965), 'Frege's theory of incomplete entities', *Philosophy of Science*, 32, pp. 329–41.

—— (1975), 'Mathematical knowledge and pattern cognition', *Canadian Journal of Philosophy*, 5, pp. 25–39.

—— (1980), *Frege and the Philosophy of Mathematics* (Ithaca, NY: Cornell University Press).

—— (1981), 'Mathematics as a science of patterns: Ontology and reference', *Nous*, 15, pp. 529–50.

—— (1982), 'Mathematics as a science of patterns: Epistemology', *Nous*, 16, pp. 95–105.

—— (1985*a*), 'How nominalist is Hartry Field's nominalism?', *Philosophical Studies*, 47, pp. 163–81.

—— (1985*b*), 'Ontology and logic: Remarks on Hartry Field's anti-platonist philosophy of mathematics', *History and Philosophy of Logic*, 6, pp. 191–209.

—— (forthcoming *a*), 'A naturalized epistemology for a Platonist mathematical ontology', to appear in *Philosophica*.

—— (forthcoming *b*), 'Beliefs about mathematical objects', to appear in A. Irvine (ed.), *Physicalism in Mathematics*.

Russell, B. (1902), 'Letter to Frege', repr. in van Heijenoort (ed.) (1967), 124–5.

—— (1906), 'On "insolubilia" and their solution by symbolic logic', repr. in his (1973), 190–214.

—— (1907), 'The regressive method of discovering the premises of mathematics', repr. in his (1973), 272–83.

—— (1919), *Introduction to Mathematical Philosophy* (London: Allen and Urwin).

—— (1973), *Essays in Analysis*, ed. D. Lacky (London: Allen and Unwin).

—— and Whitehead, A. N. (1913) *Principia Mathematica*, 3 vols. (Cambridge: Cambridge University Press).

Salmon, N. (1981), *Reference and Essence* (Princeton, NJ: Princeton University Press).

Scott, D. (1961), 'Measurable cardinals and constructible sets', *Bulletin de l'Académie Polonaise des Sciences*, 7, pp. 145–9.

—— (1977), 'Foreword', in J. L. Bell, *Boolean-Valued Models and*

Independence Proofs in Set Theory (Oxford: Oxford University Press), pp. xi–xviii.

SEARLE, J. (1958), 'Proper names', *Mind*, 67, pp. 166–73.

SHAPIRO, S. (1983*a*), 'Mathematics and reality', *Philosophy of Science*, 50, pp. 523–48.

—— (1983*b*), 'Conservativeness and incompleteness', *Journal of Philosophy*, 80, pp. 521–31.

—— (1985), 'Second-order languages and mathematical practice', *Journal of Symbolic Logic*, 50, pp. 714–42.

—— (forthcoming), 'Structure and ontology', to appear in *Philosophical Topics*.

SHOENFIELD, J. R. (1977), 'Axioms of set theory', in Barwise (ed.) (1977), 321–44.

SIERPIŃSKI, W. (1918), 'L'Axiome de M. Zermelo et son rôle dans la théorie des ensembles et l'analyse', *Bulletin de l'Académie des Sciences de Cracovie*, pp. 97–152.

—— (1924), 'Sur une propriété des ensembles ambigus', *Fundamenta Mathematicae*, 6, pp. 1–5.

—— (1925), 'Sur une classe d'ensembles', *Fundamenta Mathematicae*, 7, pp. 237–43.

—— (1934), *Hypothèse du continu* (Warsaw: Garasiński).

—— and TARSKI, A. (1930), 'Sur une propriété caractéristique des nombres inaccessibles', *Fundamenta Mathematicae*, 15, pp. 292–300.

SIMPSON, S. (1988), 'Partial realizations of Hilbert's program', *Journal of Symbolic Logic*, 53, pp. 349–63.

SKOLEM, T. (1923), 'Some remarks on axiomatized set theory', repr. in van Heijenoort (ed.) (1967), 290–301.

SKYRMS, B. (1967), 'An explication of "X knows that p"', *Journal of Philosophy*, 64, pp. 373–89.

SOLOVAY, R. M. (1969), 'On the cardinality of Σ_2^1 sets of reals', in J. Bulloff, T. Holyoke, and S. Hahn (eds.), *Foundations of Mathematics* (Berlin: Springer-Verlag), 58–73.

—— (1970), 'A model of set theory in which every set of reals is Lebesgue measurable', *Annals of Mathematics*, 92, pp. 1–56.

—— REINHARDT, W. N., and KANAMORI, A. (1978), 'Strong axioms of infinity and elementary embeddings', *Annals of Mathematical Logic*, 13, pp. 73–116.

STEINER, M. (1973), 'Platonism and the causal theory of knowledge', *Journal of Philosophy*, 70, pp. 57–66.

—— (1975*a*), *Mathematical Knowledge* (Ithaca, NY: Cornell University Press).

—— (1975*b*), 'Review of Chihara's *Ontology and the Vicious Circle Principle*', *Journal of Philosophy* 72, pp. 184–96.

STEINER, M. (1978), 'Mathematical explanation', *Philosophical Studies*, 34, pp. 135–51.

STRAWSON, P. (1959), *Individuals* (London: Methuen).

SUSLIN, M. (1917), 'Sur une définition des ensembles mesurables B sans nombres transfinis', *Comptes rendus de l'Académie des Sciences de Paris*, 164, pp. 88–91.

TARSKI, A. (1933), 'The concept of truth in formalized languages', repr. in his *Logic, Semantics, and Metamathematics,* 2nd edn. (Indianapolis, Ind.: Hackett, 1983), 152–278.

TROELSTRA, A. S. (1969), *Principles of Intuitionism* (Berlin: Springer-Verlag).

ULAM, S. (1930), 'Zur Masstheorie in der allgemeinen Mengenlehre', *Fundamenta Mathematicae*, 16, pp. 140–50.

URMSON, J. O. (1956), *Philosophical Analysis* (Oxford: Oxford University Press).

VAN HEIJENOORT, J. (ed.) (1967), *From Frege to Gödel* (Cambridge, Mass.: Harvard University Press).

VON NEUMANN, J. (1923), 'On the introduction of transfinite numbers', repr. in van Heijenoort (ed.) (1967), 346–54.

—— (1925), 'An axiomatization of set theory', repr. in van Heijenoort (ed.) (1967), 393–413.

WANG, H. (1974*a*), 'The concept of set', ch. 6 of his (1974*b*), repr. in Benacerraf and Putnam (eds.) (1983), 530–70.

—— (1974*b*), *From Mathematics to Philosophy* (London: Routledge and Kegan Paul).

WEDBERG, A. (1955), *Plato's Philosophy of Mathematics* (Stockholm: Almqvist and Wiksell).

WILLIAMSON, J. H. (1962), *Lebesgue Integration* (New York: Holt, Rinehart, and Winston).

WILSON, M. (1979), 'Maxwell's condition—Goodman's problem', *British Journal for the Philosophy of Science*, 30, pp. 107–23.

—— (1985), 'What is this thing called "pain"?—the philosophy of science behind the contemporary debate', *Pacific Philosophical Quarterly*, 66, pp. 227–67.

—— (*forthcoming*), 'Honorable intensions', in S. Wagner and R. Warner (eds.) *Notes Against a Program.*

WITTGENSTEIN, L. (1953), *Philosophical Investigations* (New York: Macmillan).

—— (1978), *Remarks on the Foundations of Mathematics* rev. edn., eds. G. H. von Wright, R. Rhees, and G. E. M. Anscombe (Cambridge, Mass.: MIT Press).

WOODIN, H. (1988), 'Supercompact cardinals, sets of reals, and weakly homogeneous trees', *Proceedings of the National Academy of Sciences (U.S.A.)*, 85, pp. 6587–91.

WRIGHT, C. (1983), *Frege's Conception of Numbers as Objects* (Aberdeen: Aberdeen University Press).

YOURGRAU, P. (1985), 'Sets, aggregates, and numbers', *Canadian Journal of Philosophy*, 15, pp. 581–92.

ZERMELO, E. (1904), 'Proof that every set can be well-ordered', repr. in van Heijenoort (ed.) (1967), 139–41.

—— (1908*a*), 'A new proof of the possibility of a well-ordering', repr. in van Heijenoort (ed.) (1967), 183–98.

—— (1908*b*), 'Investigations in the foundations of set theory, I', repr. in van Heijenoort (ed.) (1967), 199–215.

—— (1930), 'Über Grenzzahlen und Mengenbereiche', *Fundamenta Mathematicae*, 16, pp. 29–47.

INDEX